SPRINGER
LAB MANUAL

Springer Japan KK

Y. Hayashizaki   S. Watanabe (Eds.)

Restriction Landmark Genomic Scanning (RLGS)

With 62 Figures

 Springer

YOSHIHIDE HAYASHIZAKI, M. D., Ph. D.
Genome Science Laboratory, RIKEN
3-1-1 Koyadai
Tsukuba, Ibaraki 305
Japan

SACHIHIKO WATANABE, Ph. D.
Shionogi Institute for Medical Science
2-5-1 Mishima
Settsu, Osaka 566
Japan

ISBN 978-4-431-68521-0   ISBN 978-4-431-67953-0 (eBook)
DOI 10.1007/978-4-431-67953-0

Library of Congress Cataloging-in-Publication Data.
Restriction landmark genomic scanning (RLGS) / Y. Hayashizaki, S. Watanabe (eds.)
        p.   cm. – (Springer lab manual)
     Includes bibliographical references and index.

     1. Gene mapping – Laboratory manuals.   I. Hayashizaki, Y. (Yoshihide),
     1957– . II. Watanabe, S. (Sachihiko), 1939– .   III. Series.
     QH445.2.R47   1997
     572.8′633 – dc21

Printed on acid-free paper

© Springer Japan 1997
Originally published by Springer-Verlag Tokyo in 1997

Cover design: Design & Production, Heidelberg
Typesetting: Best-set Typesetter Ltd., Hong Kong

For Verne M. Chapman

Preface

In the past decade, the development of genome technology has greatly advanced toward higher speeds and the ability to handle a larger number of DNA samples. Such progress has started to change the style of research in the life sciences. One of the major advances is high-speed technology for extracting genetic information from genomic DNA and its transcripts, and the availability of large-scale sequencing information of expression sequence tags (EST) and genomic DNA. The concept of "genome scanning" that is introduced in this book arose from the need for high-speed surveying of transcripts and genomic DNA from the complex genomes of higher organisms.

The recent advances in genome data for genome mapping and sequencing have changed the style of research. For example, for identifying genes having related sequences with known genes, it is much faster to screen the EST data on a computer database rather than using traditional methods. High-speed genome scanning technology also has revolutionized research. Until now, a researcher undertaking positional cloning or other genetic analysis requiring a genome map in a species with no preexisting map faced an almost impossible task. However, genome scanning technology now provides a very powerful tool for easily obtaining the genetic map, thus facilitating the identification of genes responsible for specific phenotypes; genomic regions showing deletion, amplification, and methylation changes; and genes showing changes of expression level in different physiological states in various tissues or cells.

The genome scanning method is a technology allowing visualization of the landmarks that are distributed throughout the genome (a landmark is defined as a site on the genome that is to be visualized). Based on the type of landmark, genome scanning technology can be classified into three categories: Southern blot with DNA markers, PCR with specific primers, and restriction landmark genomic scanning (RLGS) with restriction landmarks. Among these genome scanning technologies, we originally developed RLGS as the highest-speed and most economical method for screening an enormous number of landmarks. It greatly facilitates the speed of analysis of the complex genomes in higher organisms and reduces the cost.

This book is organized into eight chapters. Chapters 1 and 2 introduce the concept of genome scanning and the principles of RLGS. Chapters 3 and 4 describe the practical protocol of the new version of the RLGS method and the method for isolating DNA clones corresponding to the locations of interest. Chapters 5–7 present the applications. Recently, restriction landmark cDNA scanning (RLCS) has been developed by which thousands of transcripts can be visualized. Chapter 8 introduces the RLCS protocol and its application.

The technology should be selected to match the purpose of the research. RLGS has a very specific character that enables the rapid scanning of a large number of landmarks located on the genome. This process facilitates construction of the complex genome map of the higher organisms and positional cloning. However, although it is useful, widely applicable, and fast, it requires specific equipment and some skill. We hope this book can provide scientists with the information necessary for efficient genome analyses using the RLGS system.

I wish to thank all authors and contributors for their indispensable efforts toward the publication of this book, and also to thank the many colleagues and secretaries for their devoted support and critical reading of the manuscript.

It is with great sadness that I must report the untimely death of Professor Verne M. Chapman, one of the most important members of the group involved in developing RLGS, who died unexpectedly in Tsukuba Science City, Japan, during a stay at The Institute of Physical and Chemical Research (RIKEN). The authors and I would like to dedicate this book to him.

Tsukuba, June 1997   YOSHIHIDE HAYASHIZAKI

Contents

List of Contributors

HAYASHIZAKI, Y.
Genome Science Laboratory
Riken Tsukuba Life Science
Center, 3-1-1 Koyadai
Tsukuba, Ibaraki 305, Japan

KAWAI, J.
Shionogi Institute for Medical
Science, Shionogi & Co., Ltd.
2-5-1 Mishima, Settsu
Osaka 566, Japan

OKAZAKI, Y.
Genome Science Laboratory
Riken Tsukuba Life Science
Center, 3-1-1 Koyadai
Tsukuba, Ibaraki 305, Japan

OHSUMI, T.
Genome Science Laboratory
Riken Tsukuba Life Science
Center, 3-1-1 Koyadai
Tsukuba, Ibaraki 305, Japan

OKUIZUMI, H.
Genome Science Laboratory
Riken Tsukuba Life Science
Center, 3-1-1 Koyadai
Tsukuba, Ibaraki 305, Japan

PLASS, C.
Molecular and Cellular
Biology Department, Roswell
Park Cancer Institute, Elm
and Carlton Streets, Buffalo,
NY 14263, USA

SASAKI, N.
Genome Science Laboratory
Riken Tsukuba Life Science
Center, 3-1-1 Koyadai
Tsukuba, Ibaraki 305, Japan

SHIBATA, H.
Genome Science Laboratory
Riken Tsukuba Life Science
Center, 3-1-1 Koyadai
Tsukuba, Ibaraki 305, Japan

SUZUKI, H.
Shionogi Institute for Medical
Science, Shionogi & Co., Ltd.
2-5-1 Mishima, Settsu
Osaka 566, Japan

TAKADA, S.
Genome Science Laboratory
Riken Tsukuba Life Science
Center, 3-1-1 Koyadai
Tsukuba, Ibaraki 305, Japan

TAKAHARA, T.
Genome Science Laboratory
Riken Tsukuba Life Science
Center, 3-1-1 Koyadai
Tsukuba, Ibaraki 305,
Japan

HELD, W. A.
Molecular and Cellular Biol-
ogy Department, Roswell Park
Cancer Institute, Elm and
Carlton Streets, Buffalo, NY
14263, USA

WATANABE, S.
Shionogi Institute for Medical
Science, Shionogi & Co., Ltd.
2-5-1, Mishima, Settsu
Osaka 566, Japan

YAOI, T.
Shionogi Institute for Medical
Science, Shionogi & Co., Ltd.
2-5-1 Mishima, Settsu
Osaka 566, Japan

Chapter 1

Concept of Genome Scanning

YOSHIHIDE HAYASHIZAKI

Genome scanning is defined as the high-speed survey of the presence or absence of landmarks throughout a genome and the measurement of their copy number in each locus. Originally, the concept of genome scanning arose from the idea of overall detection of the physical condition of whole genomic DNA. From this standpoint, it would be simplest to detect all fragments of genomic DNA generated by restriction-enzyme cleavage after electrophoresis and staining with ethidium bromide. Initially, efforts were made to visualize whole genomic DNA fragments according to this approach using the *E. coli* genome [1]. However, this approach has been limited to small-sized genomes (Fig. 1.1). As the complexity of the genome increases, the copy number of DNA molecules of the haploid genome equivalent decreases in proportion to the amount of genomic DNA. In the case of the human genome, which is 3×10^9 bp (approximately 10^3-fold of *E. coli* genome), a single copy locus per haploid genome of 1 μg human genomic DNA produces only 0.5 attomol (5×10^{-19} mol) fragments. Also, as the genome complexity increases, it becomes more difficult to separate and detect the large number of DNA fragments produced from the large-sized genome of higher organisms. Generally, to achieve the high resolution needed for DNA separation of such large genomes, the total amount of genomic DNA is technically limited, with separation steps of DNA fragments such as electrophoretic techniques.

To overcome this problem, scanning only landmark information was tried, ignoring the other genomic DNA regions. A DNA landmark would be a guidepost in the genome which can be visualized as a signal through the assay method (scanning method), representing a one-to-one correspondence of a signal to its locus. The ideal landmark and scanning method should offer the following:

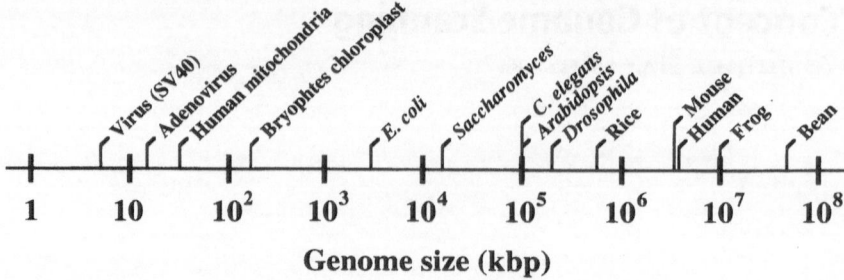

Fig. 1.1. Scale of the genome size of each organism

- robust landmark signals which can be detected reproducibly
- high-speed scanning ability (multiplex scanning)
- measurement of landmark copy number from its signal
- simple procedure and robotization
- expandability of scanning field
- applicability to any organism
- exportability of the information of landmark to any organism
- exportability of the technique and the information to other laboratories.

Originally, landmark was defined as a guidepost which represents and visualizes the locus of the genome, in its general sense. Several types of landmarks have been used for genetic analyses. Table 1.1 shows the sorts of the landmarks and the methods for visualizing them. Landmarks representing the genomic loci are classified into three categories, DNA landmarks, gene products, and phenotype.

In the process of analyzing the genome using these landmarks, the information of the chromosomal location of these landmarks is very important not only for the analysis of genomic DNA itself, but also for the identification of genes, in other words, the connection of the mutant phenotype to the responsible gene. Two types of mapping methods have been used, genetic (linkage) mapping and physical mapping. Genetic mapping is based on the principle that the distance between two loci is measured by the calculation of the recombination efficiency between paternal and maternal alleles during meiosis. The molecular basis of the

Table 1.1. Landmarks representing loci on the genome

Landmark	Method for visualizing landmark
1. DNA landmark	
DNA probe	Southern blotting
PCR (Primer)	PCR (SSLP PCR, arbitrary PCR, Alu PCR, differential display)
Restriction landmark	Restriction landmark genomic scanning, Restriction landmark cDNA scanning (RLGS, RLCS)
2. Gene products (peptides)	SDS electrophoresis, IF electrophoresis, etc.
3. Phenotype	Observation, diagnosis (Symptoms; macroscopic insights, histology)

meiotic recombination is known as chiasmatic formation (synaptonemal complex) and the principle of the calculation of the distance between two loci is that the frequency of forming the chiasma must increase in proportion to the distance. Therefore, the unit of the genetic mapping is called Morgan, and usually, we use the cM unit, which implies that one recombination event occurs during 100 meioses. The greatest advantage of this mapping method is that all the landmarks listed in Table 1.1 can be mapped based on this system, although the paternal and maternal alleles have to be discriminated in the analyses of pedigree in order to trace the transmission of each allele. On the other hand, physical maps, especially contig maps of DNA clones (the aligned clone set), provide more direct information specific to the DNA molecules. In contrast with genetic maps, the unit base pair (bp) is used with physical maps. Recently, the yeast artificial chromosome (YAC) [2], the bacterial artificial chromosome (BAC) [3], and the P1 artifical chromosome [4] have been developed, which can carry very large segments of DNA (100 kbp–1 Mbp). Using these vectors, an appropriate number of clones can cover the entire genome of higher organisms. The YAC contig clones have already been reported to cover more than 90% of the entire human genome, although they include many chimeric clones [5]. The physical map carries the DNA (gene) itself, making it a powerful resource in the location of genes. However, only DNA landmarks can be identified on this map. Therefore, for positional cloning and postitional candidate cloning [6], the location of the phenotype landmark on the physical map can be identified by the relative location of the DNA landmark to the phenotype landmark on the genetic map. DNA markers thus play a very important role as bridges to connect the genetic map, thereby providing the relative location between the phenotype landmark and the DNA landmark (only the DNA landmark can be mapped). Three sorts of DNA landmarks have been widely used, as shown in Table 1.1, DNA probes, PCR, and restriction landmarks.

Development of Southern hybridization initially provided a quantitative and qualitative method for visualizing landmarks directly from genomic DNA using a probe with a unique sequence [7]. In this system, a signal specific to each locus can be detected because characteristic molecule recognition of DNA hybridization occurs in a sequence-specific manner. Recently, polymerase chain reaction (PCR) was developed as an alternative method for detecting landmarks [8]. PCR is a technique for amplifying a region of several hundred base pairs, flanked by a pair of oligonucleotide primers. In the PCR technique, the specificity of signals depends on the specificity of the hybridization of the primers to a unique genomic locus.

To facilitate genome analysis, two approaches can be considered for identifying landmarks on complex genomes such as mammals, as shown in Fig. 1.2. One approach is simplification of the procedure for detecting loci by establishing robotized systems. In Southern hybridization using a unique probe, and in PCR for amplifying a unique sequence, a robust but single landmark can be identified. To establish hybridization- or PCR-based genome scanning, the development of a robotized system is essential (transverse axis in Fig. 1.2), because only one locus can be assayed in one procedure. The other approach is the development of a multiplex method by which multiple loci can be screened in one procedure (vertical axis in Fig. 1.2). Typical techniques of hybridization-based and PCR-based multiplex methods are DNA fingerprinting using repeating sequences (VNTR, IAP, etc.) as probes [9,10] and the Alu-PCR method [11], respectively. Trials to identify mutant loci [10] and construct genome maps [12] have proved successful in some cases, using these conventional approaches based on the concept of genome scanning. However, the scanning field and scanning speed are limited by the type of available repeating sequences, because the basic principle of both approaches to produce signal specificity is sequence-to-sequence hybridization.

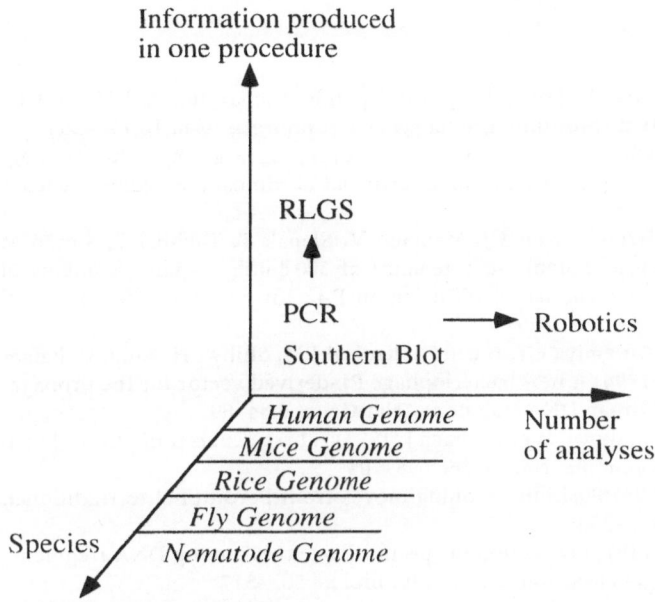

Fig. 1.2. Direction of the development of the genome scanning technology

Recently, our group introduced a new concept, termed restriction landmark, in which each restriction-enzyme recognition
site can be used as a landmark. Based on this concept, we
developed a restriction landmark genomic scanning (RLGS)
method [13–16]. Furthermore, the development of the method
for scanning restriction landmarks was extended to screen landmarks located on transcripts-restriction landmark cDNA scanning (RLCS) [17]. These two methods of restriction landmark
scanning employ (1) direct endlabeling of the genomic DNA
digested with a restriction enzyme and (2) high-resolution two-
dimensional electrophoresis. These methods provide alternative
multiplex approaches to genome analysis. Because of the strict
specificity of sequence recognition by the restriction enzyme, this
technique enables us to scan multiple and robust landmarks. In
this issue, first, we introduce the concept and principle of RLGS
and RLCS, which allow restriction landmarks to be surveyed
throughout a genome and transcripts. Second, we will introduce
a practical method which allows the detection of restriction
landmarks distributed through the large genomes of higher
eukaryotes.

References

1. Fisher SG, Lerman LS (1979) Length-independent separation of DNA restriction fragments in two-dimensional gel electrophoresis. Cell 16:191–200
2. Burke DT, Carle GF, Olson MV (1987) Cloning of large segments of exogenous DNA into yeast by means of artificial chromosome vectors. Science 236:806–812
3. Shizuya H, Birren B, Kim UJ, Mancino V, Slepak T, Tachiiri Y, Simon M (1992) Cloning and stable maintenance of 300-kilobase-pair fragments of human DNA in *Escherichia coli* using an F-factor-based vector. Proc Natl Acad Sci USA 89:8794–8797
4. Ioannou PA, Amemiya CT, Garnes J, Kroisel PM, Shibya H, Chen C, Batzer MA, Jong PJ (1994) A new bacteriophage P1-derived vector for the propagation of large human DNA fragments. Nat Genet 6:84–89
5. Cohen D, Chumakov I, Weissenbach J (1993) A first-generation physical map of the human genome. Nature 366:698–701
6. Collins FS (1995) Positional cloning moves from perdition to traditional. Nat Genet 9:347–350
7. Southern EM (1975) Detection of specific sequences among DNA fragments separated by gel electrophoresis. J Mol Biol 98:503–517
8. Saiki RK, Scharf S, Faloona F, Mullis KB, Horn GT, Erlich HA, Arnheim N (1985) Enzymatic amplification of β-globin genomic sequences and restriction site analysis for diagnosis of sickle cell anemia. Science 230:1350–1354
9. Uitterlinden AG, Slagboom PE, Knook DL, Vijg J (1989) Two dimensional DNA fingerprinting of human individuals. Proc Natl Acad Sci USA 86:2742–2746
10. Brilliant MH, Gondo Y, Eicher EM (1991) Direct molecular identification of the mouse pink-eyed unstable mutation by genome scanning. Science 252:566–569
11. Nelson DL, Ledbetter SA, Corbo L, Victoria MF, Ramirez-Slis R, Webster TD, Ledbetter DH, Caskey CT (1989) Alu polymerase chain reaction: a method for rapid isolation of human-specific sequences from complex DNA sources. Proc Natl Acad Sci USA 86:6686–6690
12. Dietrich W, Katz H, Lincoln SE, Shin H-S, Friedman J, Dracopoli NC, Lander ES (1992) A genetic map of the mouse suitable for typing intraspecific crosses. Genetics 131:423–447
13. Hayashizaki Y, Hirotsune S, Okazaki Y, Muramatsu M, Asakawa J (1994) Fundamentals and applications: restriction landmark genomic scanning (RLGS). In: Meyers RA (ed) The encyclopedia of molecular biology. VCH, Weinheim, pp 813–817
14. Hatade I, Hayashizaki Y, Hirotsune S, Komatsubara H, Mukai TA (1991) Genomic scanning method of higher organisms using restriction sites as landmarks. Proc Natl Acad Sci USA 88:9523–9527
15. Hayashizaki Y, Hirotsune S, Okazaki Y, Hatada I, Shibata H, Kawai J, Hirose K, Watanabe S, Fushiki S, Wada S, Sugimoto T, Kobayakawa K, Kawara T, Sibuya T, Mukai T (1993) Restriction landmark genomic scanning method and its various applications. Electrophoresis 14:251–258

16. Okazaki Y, Okuizumi S, Sasaki N, Ohsumi T, Kuromitsu J, Kataoka H, Muramatsu M, Iwadate A, Hirota N, Kitajima M, Plass C, Chapman VM, Hayashizaki Y (1994) A genetic linkage map of the mouse using an expanded production system of restriction landmark genomic scanning (RLGS Ver.1.8). Biochem Biophys Res Commun 205:1922–1929
17. Suzuki H, Yaoi T, Kawai J, Hara A, Kuwajima G, Watanabe S (1996) Restriction landmark cDNA scanning (RLCS): a novel cDNA display system using two-dimensional gel electrophoresis. Nucleic Acids Res 24:289–294

Chapter 2

Principle of RLGS

Yoshihide Hayashizaki

Contents

2.1
Breakthroughs Enabling the Development of RLGS

Direct endlabeling followed by electrophoretic separation of mammalian genomic DNA has been very difficult to perform because of its high complexity. However, three significant breakthroughs have led to the development of RLGS. First, effective restriction endonuclease for genomic analysis has been discovered. Restriction enzymes which recognize 4- or 6-base pair (bp) sequences, conventionally used produce so many DNA fragments that these fragments cannot be separated as discrete signals by electroporesis. Recently, various enzymes have been discovered which recognize 8-bp or rare sequences and produce an appropriate number of DNA fragments for analysis.

Second, a high-resolution electrophoretic system was established. Methods for thin-layer slab agarose gel or fine disk agarose gel electrophoreses for the first dimension (1st-D) have been developed to enable high resolutional electrophoretic separation [1–3]. This enabled us to subject the 1st-D agarose samples to high-resolution polyacrylamide gel electrophoresis in the second dimension (2nd-D), because the thickness of the second polyacrylamide gel depends on the thickness of the first agarose gel.

Third, a technique to reduce nonspecific signals which arise from incorporation of radioactivity into damaged sites of genomic DNA has been developed. As the genome size increases, the copy number of DNA fragments generated decreases proportionally.

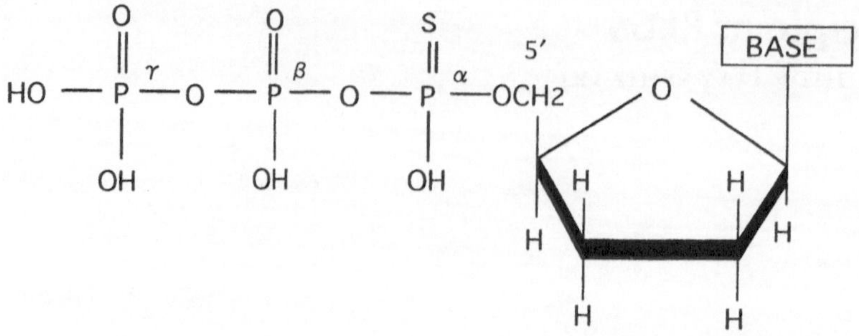

Fig. 2.1. Dideoxy nucleotide analog for the blocking method (dideoxynucleoside [α-thio] triphosphate)

However, the background produced by nonspecific incorporation of radioactivity does not change. Therefore, the signal-to-noise ratio decreases, depending on the genome size. Generally, when the genome size increases to more than 1×10^8 bp, specifc signals are very difficult to detect because of the background interference.

To overcome this problem, we developed a blocking technique to reduce the nonspecific incorporation of radioactivity. The principle of the blocking technique is based on the incorporation of the dideoxy nucleotide analog into damaged sites such as nicks, gaps, and/or double-strand breaks. After the blocking reaction, the DNA fragment cannot be elongated by DNA polymerase activity from these sites containing nucleotide analogs. Moreover, these analogs, such as dideoxy-[α-thio] nucleotides, substitute a sulfur atom in place of the oxygen atom and thereby interrupt the 3′ exonuclease activity of DNA polymerase in the labeling step (Fig. 2.1).

2.2
Principle and Procedure of RLGS

RLGS is a technique for scanning restriction landmarks at intervals of less than 1 Mb. As mentioned above, the random cleavage caused during DNA preparation affects the RLGS pattern. To produce a clear pattern with low background, DNA of high quality is essential. The critical factors are not only the average length and the extent of the damage (nicks and gaps), but also the fraction of the DNA population less than 10 kbp in length (such small DNA fragments increase the background noise). In the DNA preparation procedure, more care should be taken to avoid enzymatic degradation (with endonuclease leaking from lysosome and other enzymes in the lysis step), rather than mechanical shearing. Therefore, when DNA is prepared by Proteinase K and the phenol extraction method, the samples should be treated for a very short time using an excess of Proteinase K.

Figure 2.2 shows the entire RLGS procedure. It consists of the blocking step (Fig. 2.2a) and the process (Fig. 2.2b–f) toward high-resolution two-dimensional electrophoresis to separate large numbers of DNA fragments very precisely.

Figure 2.2a. Blocking. The principle of the blocking procedure is mentioned above. Blocking is performed in order to reduce the background generated by the incorporation of radioactivity into the nonspecifically damaged sites.

Figure 2.2b. DNA digestion with restriction enzyme E_L. This step comprises the cleavage of genomic DNA by a restriction enzyme. The resulting restriction sites are then used as landmarks. The type of restriction enzyme used, however, should be chosen according to the frequency of its recognition sites in the genome and its sensitivity to DNA methylation. DNA methylation affects the RLGS pattern when a methylation-sensitive enzyme is used as restriction enzyme E_L. In other words, the change in DNA methylation throughout the genome can be scanned using a methylation-sensitive enzyme.

Figure 2.2c. Labeling. The labeling method depends on the shape of the restriction site. For a 5′-protruding end, a filling reaction of [α-^{32}P] deoxynucleotide with Sequenase Ver. 2 is used. For a 3′-protruding (or blunt) end, terminal deoxynucleotidyl transferase (terminal transferase) is used, which can corporate [α-^{32}P] dideoxynucleotides on the 3′ termini of DNA fragments.

Figure 2.2d. Fragmentation of the labeled DNA with restriction enzyme E_B. This step comprises the cleavage of the genomic DNA radiolabeled at the recognition sites of restriction enzyme E_L. Because the length of human and other mammalian genomes is approximately 3×10^9 bp, more than 3000 DNA fragments with an average length of 1 Mbp will be produced, even if NotI, the most infrequent 8-bp recognition enzyme, is used as restriction enzyme E_L. However, such long and numerous DNA fragments cannot be separated by conventional methods such as pulse field gel electrophoresis (PFGE). To solve this problem, the digestion of radiolabeled fragments with restriction enzyme E_B (its recognition

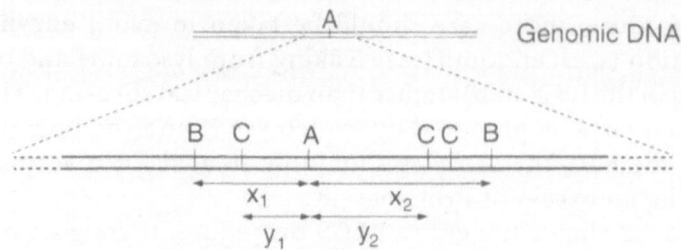

A: Site for restriction enzyme E_L (Restriction landmark)
B: Site for restriction enzyme E_B
C: Site for restriction enzyme E_C

(a) Blocking (with analogues of nucleotide)
(b) Landmark cleavage (with restriction enzyme E_L)
(c) Labeling (at restriction landmark)   ①②: Isotope label

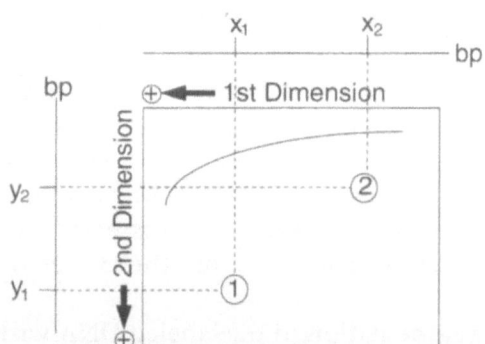

(d) The fragmentation with restriction enzyme E_B
(e) The first fractionation (by 0.8% agarose gel electrophoresis)
(f) The fragmentation of labeled DNA with restriction enzyme E_C
(g) The second fractionation
     (by 5% polyacrylamide gel electrophoresis)
(h) Autoradiography

Fig. 2.2. Procedure for genome scanning by two-dimensional gel electrophoresis. x_1 and x_2 represent the distance from a restriction landmark to the neighboring site for restriction enzymes E_L or E_B. y_1 and y_2 represent the distance from the E_L site to the E_C site. ① and ② indicate the labeled sites at the end of the landmark [8]. (By permission of VCH)

sites appearing more frequently in the genome than restriction landmarks) is employed. So that only a workable number of spots (less than 3000 spots in one gel) appear on the RLGS profile, an appropriate restriction enzynme, E_B, should be selected.

Reduction in the length of the DNA fragments enables the use of the high-resolution thin-layer agarose gel or disk agarose gel, in which the entire resolution range is expanded to more than 40 cm.

In the case of the number of restriction sites resulting from the use of restriction enzyme E_L being greater than 6000 (the average length of the DNA fragments being less than 500 kbp), the treatment of restriction enzyme E_B, step d (Fig. 2.2d), is sometimes omitted. Although this number of spots cannot be separated in one gel using the present technique, these fragments will be trapped at the top of the agarose gel, resulting in the reduction of the spot number on an RLGS profile. This is because the average length of DNA fragments digested with almost all 8-bp cutters and rare cutters will be more than 30 kbp. An appropriate number of spots dispersed over a two-dimensional field is required for the final RLGS profile.

Figure 2.2e. First fractionation by agarose gel electrophoresis. For the first-dimensional (1st-D) agarose gel electrophoresis, two methods are availabe. One uses a horizontal gel and the other a vertical gel. For both, in order to obtain a good RLGS profile, the agarose gel should be as thin and fine as possible and the samples should be electrophoresed long enough to precisely separate more than 3000 signals.

Figure 2.2f. Fragmentation of labeled DNA with restriction enzyme E_C. To resolve these numerous signals, fragments are subjected to yet further separation. The agarose gel strip or gel cylinder is treated with a reaction mixture containing restriction enzyme E_C. Usually, an enzyme whose restriction sites are more frequent than those of restriction enzyme E_B, such as a 4-bp cutter, should be used as restriction enzyme E_C. This cleavage reaction causes the DNA fragments to differ in electrophoretical mobility depending on the distance from the site of restriction enzyme E_L to E_C (y1 and y2 in Fig. 2.2).

Figure 2.2g. Second fractionation. The DNA fragments are subjected to the second-dimensional (2nd-D) polyacrylamide gel electrophoresis by connection of the agarose strip to the 2nd gel. The sizes of the DNA fragments range from 2 kbp to 70 bp. Technically, horizontal gel or vertical gels can be used alternatively.

Figure 2.2h. Autoradiography. The final gel samples are dried and autoradiographed.

2.3
Advantages of RLGS

Figure 2.3 shows the resulting RLGS pattern, using *Not*I – *Eco*RV – *Mbo*I, as restriction enzymes E_L–E_B–E_C, respectively. The resulting X, Y coordinates of each spot indicate the distance between the restriction landmark (restriction enzyme E_L) and the site of restriction enzymes E_B and E_C, respectively (Fig. 2.2). On the whole, in the complex genome of higher organisms, even 8-bp recognition enzymes produce more than 1000 fragments. Although there is such a large number of spots/loci, they could be completely separated by the high-resolution two-dimensional separation which uses the two parameters X and Y.

Thus, a spot on a RLGS profile corresponds to a unique locus even in a large genome, resulting in a functional landmark. RLGS has the following advantages for genome scanning:

- It has high-speed scanning capability. Thousands of restriction landmarks can be scanned simultaneously.

- Using alternate enzymes as restriction enzyme E_L increases in the number of landmarks (Table 2.1).

- This method can be applied to any organism, including mammals and plants, because direct labeling of restriction sites (not hybridization procedure) is employed as the detection system.

- Spot intensity reflects the copy number of the DNA fragment. In Fig. 2.3, several enhanced spots can be seen. These spots are derived from repetitive sequences such as the encoding sequence of ribosomes. A haploid genomic fragment can be distinguished from a diploid fragment by its signal intensity on an RLGS profile.

- Using a methylation-sensitive enzyme, the methylation state of genomic DNA can be screened.

- A CpG island, which is abundant in GC bases and is thought to be a transcription regulatory region, is usually present within 2 kb of the 5′ transcriptional end of a functional gene. For example, it is known that 89% of *Not*I recognition sites are located on CpG islands [4]. CpG-rich enzymes such as *Not*I or *Bss*HII are preferentially used as restriction enzyme E_L of RLGS

Table 2.1. Enzymes sensitive and insensitive to vertebrate methylation [8]

^{5Me}CG-sensitive enzyme		^{5Me}CG-insensitive enzyme	
NotI	5'GCGGCCGC3'	PacI	5'AATTAATT3'
SfiI	5'GGCCNNNNNGGCC3'	SwaI	5'ATTTAAAT3'
FseI	5'GGCCGGCC3'	Sse8387I	5'CCTGCAGG3'
CspI	5'CGG(A/T)CCG3'	PmeI	5'GTTTAAAC3'
MluI	5'ACGCGT3'		
AscI	5'GGCGCGCC3'		
BssHII	5'GCGCGC3'		
ClaI	5'ATCGAT3'		
XhoI	5'CTCGAG3'		
Eco52I	5'CGGCCG3'		
SalI	5'GTCGAC3'		
NarI	5'GGCGCC3'		
NruI	5'TCGCGA3'		
RsrII	5'CGG(A/T)CCG3'		

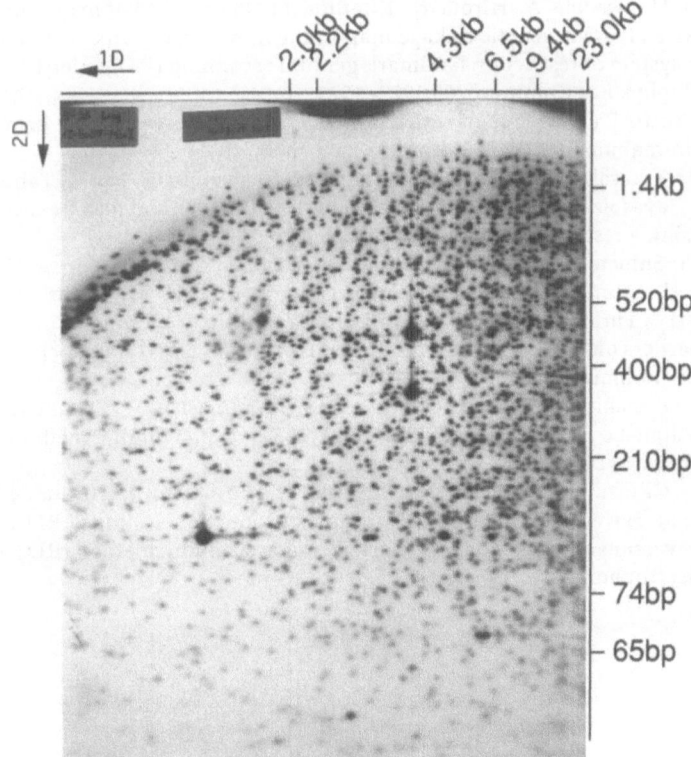

Fig. 2.3. RLGS profile using NotI-EcoRV-MboI as restriction enzymes, E_L-E_B-E_C

procedure [2,4,5]. Thus, RLGS using CpG-rich enzymes is an effective method for the search for and investigation of functional genes.

- The DNA fragments corresponding to specific spots can be cloned from the DNA eluted from the punched-out get [6,7]. The clone is useful as a probe for library screening or as a sequence tagged site (STS) marker after sequencing.

References

1. Hatada I, Hayashizaki Y, Hirotsune S, Komatsubara K, Mukai T (1991) A genomic scanning method for higher organisms using restriction sites as landmarks on the genome. Proc Natl Acad Sci USA 88:9523-9527
2. Hayashizaki Y, Hirotsune S, Okazaki Y, Hatada I, Shibata H, Kawai J, Hirose K, Watanabe S, Fushiki S, Wada S, Sugimoto T, Kobayakawa K, Kawara T, Katsuki M, Sibuya T, Mukai T (1993) Restriction landmark genomic scanning and its various applications. Electrophoresis 14:251-258
3. Okazaki Y, Okuizumi H, Sasaki N, Ohsumi T, Kuromitsu J, Kataoka H, Muramatsu M, Iwadate A, Hirota N, Kitajima M, Plass C, Chapman VM, Hayashizaki Y (1994) A genetic linkage map of the mouse using an expanded production system of restriction landmark genomic scanning (RLGS Ver.1.8). Biochem Biophys Res Commun 205:1922-1929
4. Lindsay S, Bird AP (1987) Use of restriction enzyme to detect potential gene sequences in mammalian DNA. Nature 327:336-338
5. Okuizumi H, Okazaki Y, Sasaki N, Muramatsu M, Nakashima K, Fan K, Tano H, Ohba K, Hayashizaki Y (1994) Application of the RLGS method to large-size genomes using a restriction trapper. DNA Res 1:99-102
6. Hirotsune S, Shibata H, Okazaki Y, Sugino H, Imoto H, Sasaki N, Hirose K, Okuizumi H, Muramatsu M, Plass C, Chapman VM, Miyamoto C, Tamatsukuri S, Furuichi Y, Hayashizaki Y (1993) Molecular cloning of polymorphic markers on RLGS gel using the spot target cloning method. Biochem Biophys Res Commun 194:1406-1412
7. Ohsumi T, Okazaki Y, Shibata H, Hirotsune S, Muramastsu M, Suzuki H, Taga C, Watanabe S, Hayashizaki Y (1995) RLGS spot cloning method. Electrophoresis 16:203-209
8. Okazaki Y, Okuizumi H, Sasaki N, Ohsumi T, Kuromitsu J, Hirota N, Muramatsu M, Hayashizaki Y (1995) A multiplex production systme of RLGS gels – a new version of restriction landmark genomic scanning method (RLGS Ver.1.8). Electrophoresis 16:197-202

Chapter 3

Protocols for RLGS Gel Production

Yasushi Okazaki, Hisato Okuizumi, Nobuya Sasaki,
Shuji Takada, Tokuei Takahara, and
Yoshihide Hayashizaki

Contents

RLGS can be performed using various enzyme combinations (restriction enzymes E_L-E_B-E_C). The labeling method must be chosen according to the shape of the recognition site of enzyme E_L. In addition, the reaction buffers used in each process should be changed, accompanied by variation of the enzyme. Here, protocols for two good representative enzyme combinations, combination 1 (*NotI-PvuII-PstI*) and combination 2 (*PacI-EcoRV-MboI*), are shown.

3.1
Preparation of Genomic DNA for RLGS

- Proteinase K (Merck, Darmstadt, Germany)   **Materials**

- RNase A (Boehringer, Mannheim, Germany)

- phenol (distilled)

- chloroform

- isoamyl alcohol

- SDS (sodium dodecyl sulfate)

- 8-hydroxyquinoline

- 50-ml Falcon tube (Becton Dickinson New Jersey, USA)

- mortar

- pestle

- dialysis tube

- liquid nitrogen (LN_2)

- spatula

- aluminum foil

Stock buffers
- TE (10 mM Tris-HCl pH 7.5, 1 mM EDTA)

- PCI (phenol-chloroform-isoamyl alcohol) [50:49:1 distilled phenol/chloroform/isoamyl alcohol, 0.1% 8-hydroxyquinoline, PCI should be buffered with TE.]

- lysis buffer (150 mM EDTA, 1% SDS, 10 mM Tris-HCl pH 8)

- Proteinase K [10 mg/ml, Proteinase K should be mixed just before use.]

- Phosphate buffered saline (PBS)

3.1.1
Preparation of Genomic DNA for RLGS from Tissue Samples

Protocol
1. Prechill the mortar, pestle, spatula, and aluminum foil in LN_2.

2. Add 2 ml of lysis buffer to 0.5 g of tissue and freeze in LN_2.

3. Wrap the tissue in aluminum foil.

4. Freeze in LN_2.

5. Crush the frozen tissue with a hammer.

6. Grind down the tissue pieces into powder using mortar and pestle.

7. Transfer the tissue powder to 50-ml Falcon tube with prechilled spatula.

8. Add 25 ml of lysis buffer with 150 μl of 10 mg/ml Proteinase K.

9. Stir gently but quickly with spatula at room temperature.

10. Incubate for 20 min at 55 °C.

11. Cool on ice for 5 min.

12. Add 25 ml of PCI.

13. Mix by rotating (25 rpm, Taitec: Rotator RT-50) for 30 min.

14. Centrifuge at 3000 rpm for 30 min.

15. Transfer the aqueous (upper) phase to a new Falcon tube with a cutoff Eppendorf tip (1000 μl).

16. Repeat PCI extraction.

17. Transfer the aqueous phase to a dialysis tube.

18. Dialysis three times against 1 l of 10 mM Tris-HCl pH 8 overnight.

19. Transfer from the dialysis tube to a new Falcon tube.

20. Add 20–40 μl of 1 mg/ml RNase A
    (final concentration of 1 μg/ml).

21. Incubate for 2 h at 37 °C.

22. Divide the sample into two Falcon tubes (about 15 ml each).

23. Add 35 ml of cold 100% ethanol to each tube.

24. Rotate gently (15 rpm, Taitec: Rotator RT-50) for 30–60 min.

25. Pick up the pellet of DNA and transfer into an Eppendorf sample tube.

26. Add 300–500 μl TE buffer (**Note:** Do not allow the pellet of DNA to dry completely, otherwise it will be very difficult to dissolve!).

27. Dissolve by tapping gently.

28. Measure DNA concentration.

3.1.2
Preparation of Genomic DNA for RLGS from Cell Samples

Protocol
1. Harvest 2×10^7 cells in a 50-ml Falcon tube.

2. Wash with 10–20 ml of PBS (–).

3. Centrifuge at 1000 rpm for 5 min at 4 °C.

4. Discard the supernatant.

5. Resuspend the pellet in 400 μl of PBS (–).

6. Add 4 ml of lysis buffer with 24 μl of 10 mg/ml Proteinase K and mix gently.

7. Incubate for 20 min at 55 °C.

8. Cool on ice for 5 min.

9. Add 4.5 ml of PCI.

10. Mix by rotating (25 rpm, Taitec: Rotator RT-50) for 30 min.

11. Centrifuge at 3000 rpm for 30 min.

12. Transfer the aqueous (upper) phase to a new Falcon tube with a cutoff Eppendorf tip (1000 μl).

13. Repeat PCI extraction.

14. Transfer the aqueous phase to the dialysis tube.

15. Dialyze three times against 1 l of 10 mM Tris-HCl pH 8 overnight.

16. Transfer from the dialysis tube to a new Falcon tube.

17. Add 5–10 μl of 1 mg/ml RNase A (final concentration 1 μg/ml).

18. Incubate for 2 h at 37 °C.

19. Add 35 ml of cold 100% ethanol.

20. Rotate gently (15 rpm, Taitec: Rotator RT-50) for 30–60 min.

21. Pick up the pellet of DNA and transfer into an Eppendorf sample tube.

22. Add 50 μl TE buffer (**Note:** Do not allow the pellet of DNA to dry completely; otherwise it will be very difficult to dissolve!).

23. Dissolve by tapping gently.

24. Measure DNA concentration.

3.2
Protocol for Combination 1 (*NotI-PvuII-PstI*)

3.2.1
Restriction Enzyme Digestion and Labeling

- DNA polymerase I (Toyobo)   **Materials**

- deoxynucleoside [α-thio] triphosphates (dGTP[α]S, dCTP[α]S; Toyobo)

- dideoxynucleoside triphosphates (ddGTP, ddATP, ddTTP, ddCTP; Toyobo)

- [α-^{32}P] dGTP (3000 Ci/mmol; Amersham, UK)

- [α-^{32}P] dCTP (6000 Ci/mmol; Amersham UK)

- *Not*I (Takara, Kyoto, Japan)

- *Pvu*II (Toyobo)

- *Sca*I (Takara)

- bovine serum albumin (BSA) (Sigma fraction V)

- Triton X-100

- Sequenase Ver. 2.0 (USB, Cleveland, USA)

- bromophenol blue (BPB)

- xylene cyanol (XC)

- pBlueScript II (pBSII) (Stratagene, San Diego, USA)

- dithiothreitol (DTT)

- salmon sperm DNA solution (500 µg/ml sonicated salmon sperm DNA in TE buffer)

- ice-cold TCA (20%)

- glass microfiber filters (2.5-cm diameter, Whatman GF/F)

- filtration device

- scintillation vials

- Scotch 3M tape (Scotch, catalog no.193J)

- Kimwipe (Kimberly-Clark Corp., Tokyo, Japan)

Stock - 10× high buffer (HB) [500 mM Tris-Cl (pH 7.4), 100 mM MgCl$_2$,
buffers   1 M NaCl, 10 mM DTT]

- 10× medium buffer (MB) [500 mM Tris-Cl (pH 7.4), 100 mM MgCl$_2$, 500 mM NaCl, 10 mM DTT]

- TE buffer

- 20× SH buffer [3M NaCl, 0.2% Triton X-100, 0.2% BSA]

- 6× dye solution [0.25% BPB, 0.25% XC in 150 mM EDTA with 30% glycerol in water]

Protocol Blocking

Blocking 1. Mix the following reagents

Sample (0.5–1.5 µg/µl DNA)	7 µl
10× HB	1 µl
1 M DTT	0.1 µl
10 mM dGTP[α]S	0.4 µl
10 mM dCTP[α]S	0.2 µl
10 mM ddATP	0.4 µl
10 mM ddTTP	0.4 µl
DNA polymerase I (3.5 U/µl)	0.5 µl
Total	10 µl

All solutions except for sample and DNA polymerase I may be premixed and stored at −20 °C as blocking buffer. Usage of blocking buffer is as follows:

Sample	7 µl
Blocking buffer	2.5 µl
DNA polymerase I (3.5 U/µl)	0.5 µl
Total	10 µl

2. Mix by pipetting.

3. Incubate for 20 min at 37 °C.

4. Incubate for 30 min at 65 °C.

*Not*I Digestion

5. Mix the following reagents.

Sample	10 µl
2.5× SH buffer	8 µl
*Not*I (10 U/µl)	2 µl
Total	20 µl

6. Mix gently but completely by pipetting

18 µl

2 µl
Control: Use 2 µl of the
mixture for cut check
Add 8 µl of 0.1 µg/8 µl pBSII/
*Sca*I cut in HB
Mix by pipetting

7. Incubate for 2 h at 37 °C

Incubate for 2 h at 37 °C
0.8% agarose
electrophoresis
(If DNA is completely cut,
two bands are seen)

Labeling

8. Mix the following Agents   **Labeling**

Sample	18 µl
1 M DTT	0.3 µl
[α-^{32}P]dGTP	1 µl
[α-^{32}P]dCTP	1 µl
Sequenase Ver. 2.0 (13 U/µl)	0.1 µl
Total	20.4 µl

9. Mix by pipetting.

10. Incubate for 30 min at 37 °C.

*Pvu*II digestion

11. Mix the following agents.

Sample	20.4 µl
1 mM ddGTP	1 µl
1 mM ddCTP	1 µl
100 mM MgCl$_2$	1.2 µl
DW	4.4 µl
PvuII (10 U/µl)	2 µl
Total	**30 µl**

All solutions except for sample and PvuII may be premixed and stored at −20°C as second digestion buffer. Usage of second digestion buffer is as follows:

Sample	20.4 µl
2nd digestion buffer	7.6 µl
PvuII (10 U/µl)	2 µl
Total	**30 µl**

12. Mix gently but completely by pipetting (30 µl).

28 µl

2 µl
Control: Use 2 µl of the mixture for cut check
Add 8 µl of 0.1 µg/8 µl pBSII/ScaI cut in HB
Mix by pipetting

13. Incubate at 37°C for 1 h

Incubate for 1 h at 37°C
0.8% agarose electrophoresis
(If DNA is completely cut, two bands are seen)

TCA Precipitation

Precipitation 14. Pick up 2 µl of labeled sample to a disposable sampling tube containing 100 µl of salmon sperm DNA solution.

15. Mix with voltex.

16. Add 100 µl of ice-cold 20% TCA.

17. Put on ice for 5 min.

18. Collect the precipitate by filtering the solution through a glass microfiber filter.

19. Wash the filter with 9 ml of ice-cold 20% TCA.

20. Transfer the filter into a scintillation vial.

21. Measure the radioactivity in a liquid scintillation counter.

22. Adjust the radioactivity to 18 000 cpm/2 μl by adding 6× dye solution (final concentration: 1.5×) and TE.

3.2.2
Two-Dimensional Separation with Electrophoresis and in Situ Restriction Enzyme Digestion

- SeaKem GTG agarose   **Materials**

- acrylamide (Daiichi Pure Chem, Tokyo, Japan)

- methylenebisacrylamide (Nakalai Tesque, Kyoto, Japan)

- ammonium persulfate(Nakalai Tesque)

- BSA (Sigma fraction V)

- disposable plastic syringes (1 ml, 6 ml)

- needles (10 cm long 19-gauge 90° cut), three-way stopcocks for syringes

- support stand with double-buret clamps

- 10% ammonium persulfate (Nakalai Tesque), N,N,N′,N′-tetramethylene ethylene-diamine (TEMED; Nakalai Tesque)

- PstI (Takara)

- 10× 1st-D buffer (for 1st-D; dissolve 242 g of Tris, 109 g of so-   **Stock**
  dium acetate trihydrate, 42 g of NaCl, 23.4 g of EDTANa$_2$ in 1.8 l   **buffers**
  of H$_2$O and adjust the pH to 8.15 with acetic acid. Adjust the
  volume to 2 l and pass through a 0.2-μm filter or autoclave. The
  pH of the 1× buffer is 8)

- 10× TBE buffer (for 2nd-D; dissolve 1.210 g Tris, 770 g boric acid, and 75 g EDTANa$_2$ in H$_2$O and adjust to 20 l)

- Dye solution (0.25% BPB, 0.25% XC in TE without glycerol)

Electrophoretic apparatus   All electrophoretic apparatuses are manufactured by Bio Craft (Tokyo, Japan) (Fig. 3.1)

- 1st-D electrophoretic tank (including tank anodal top and cathodal bottom)
- 1st-D gel holder (including Teflon tubing)
- Silicon stopper
- 2nd-D electrophoretic tank (Fig. 3.2a,b)
- 2nd-D electrophoretic plate
- 2nd-D electrophoretic spacer

Protocol   **1st-D Gel Casting**

Gel casting   1. Mix the following agents and adjust the volume to 30 ml with DW.

Fig. 3.1. Electrophoretic apparatus for RLGS. Original method of 1st-D electrophoresis using Teflon tubing capillary was developed by Dr. Asakawa [1,2], and modified at RIKEN as shown in this figure. (Photo Biocraft catalog)

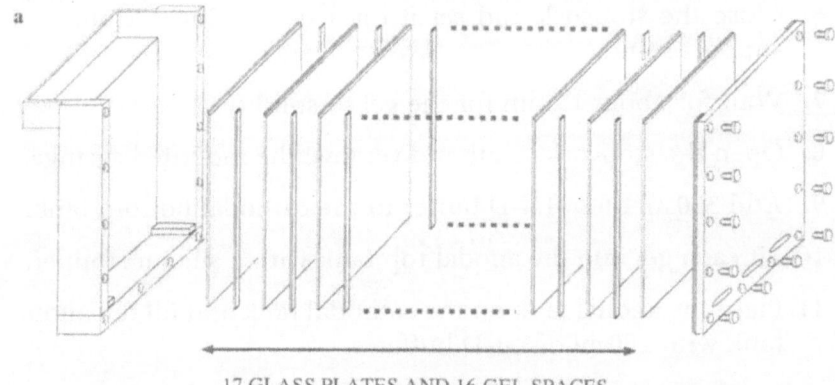

17 GLASS PLATES AND 16 GEL SPACES

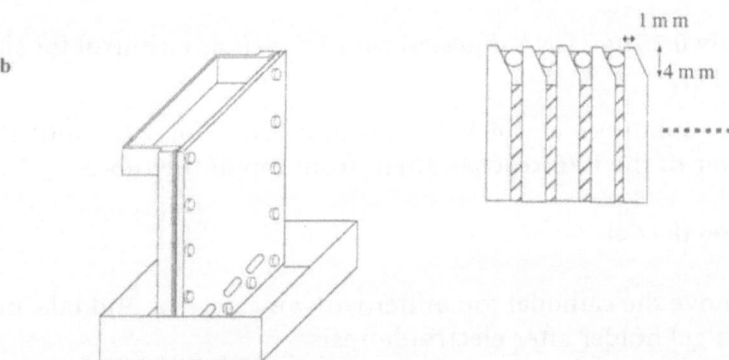

1 mm

4 mm

Fig. 3.2. a Assembly of glass plates. **b** Electrophoretic apparatus and side view of the connection of 1st agarose gel noodle to 2nd polyacrylamide gel [3]. (By permission of Academic Press)

SeaKem GTG agarose	0.24 g
10× 1st-D buffer	3 ml
60% sucrose	2.5 ml

2. Melt agarose in a microwave oven.

3. Cool the solution to 55–65 °C.

4. Connect a 5-ml plastic syringe fitted with a three-way stopcock and the gel holder with 3 cm of silicon tubing.

5. Suck up the gel solution gradually to reach the line (1 cm below the top).

6. Close the stopcock and set it on a double buret clamp on a support stand.

7. Wait for about 15 min for the gel to solidify.

8. Open the stopcock to air and remove the mounted syringe.

9. Add 350 ml of 1× 1st-D buffer to the cathodal bottom tank.

10. Fit each gel into the anodal top tank with a silicon stopper.

11. Place the anodal tank on the cathodal tank and fill the anondal tank with 250 ml of 1st-D buffer.

Running the 1st-D Gel

Gel running

12. Apply 15 μl of sample.

13. Apply 0.15 μg λDNA digested with *Hind*III as a control for size marker.

14. Electrophoresis at 100 V for 2 h and 230 V for 24 h until the center of the BPB reaches 50 cm from top of the tube.

Extruding the Gel

Gel extruding

15. Remove the cathodal top buffer with an aspirator and take out each gel holder after electrophoresis.

16. Expel the gel slowly using a cut-off yellow Eppendorf tip on a 1-ml syringe (Fig. 3.3).

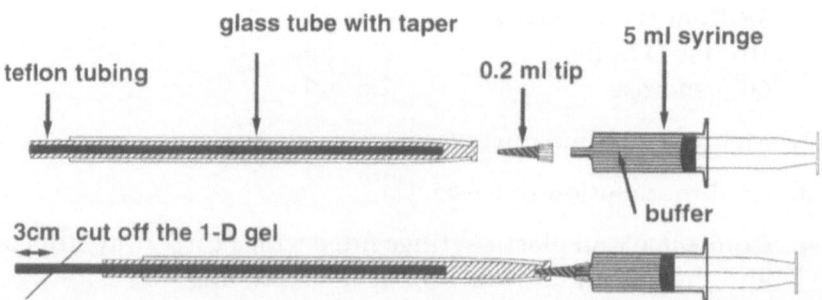

Fig. 3.3. RLGS profile using *Not*I-*Eco*RV-*Mbo*I as restriction enzymes E_L-E_B-E_C. Original method of 1st-D electrophoresis using Teflon tubing capillary was developed by Dr. Asakawa [1,2], and modified at RIKEN as shown in this figure

17. Cut off the point of 500 bp and discard the expelled gel (**Caution:** it is radioactive!)

18. Soak the gel in 50-ml Falcon tube containing 40 ml of 1× HB (Fig. 3.4).

19. Rock the tubes gently and equilibrate the gel for 10 min.

20. Change the buffer and equilibrate once again for 10 min.

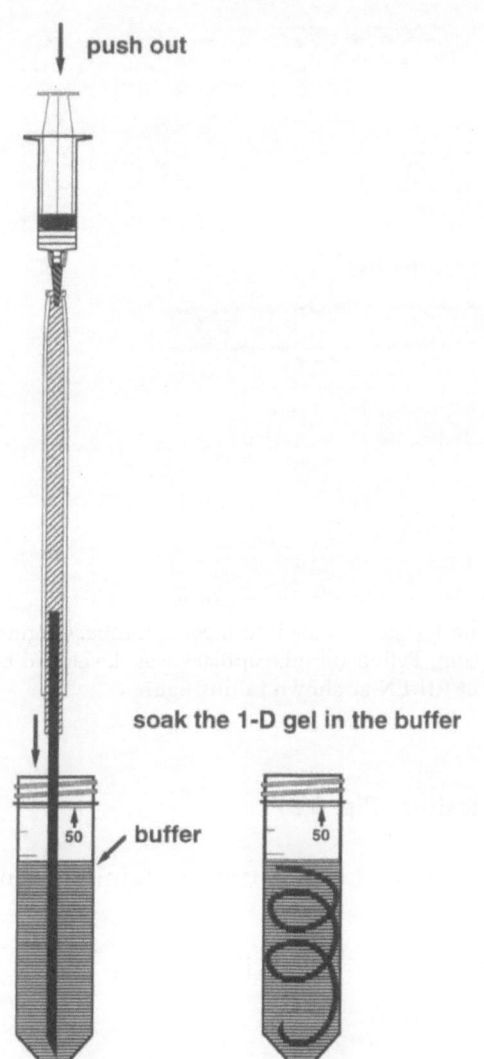

Fig. 3.4. Dialysis of 1-D agarose gel noodle. Original method of 1st-D electrophoresis using Teflon tubing capillary was developed by Dr. Asakawa [1,2], and modified at RIKEN as shown in this figure

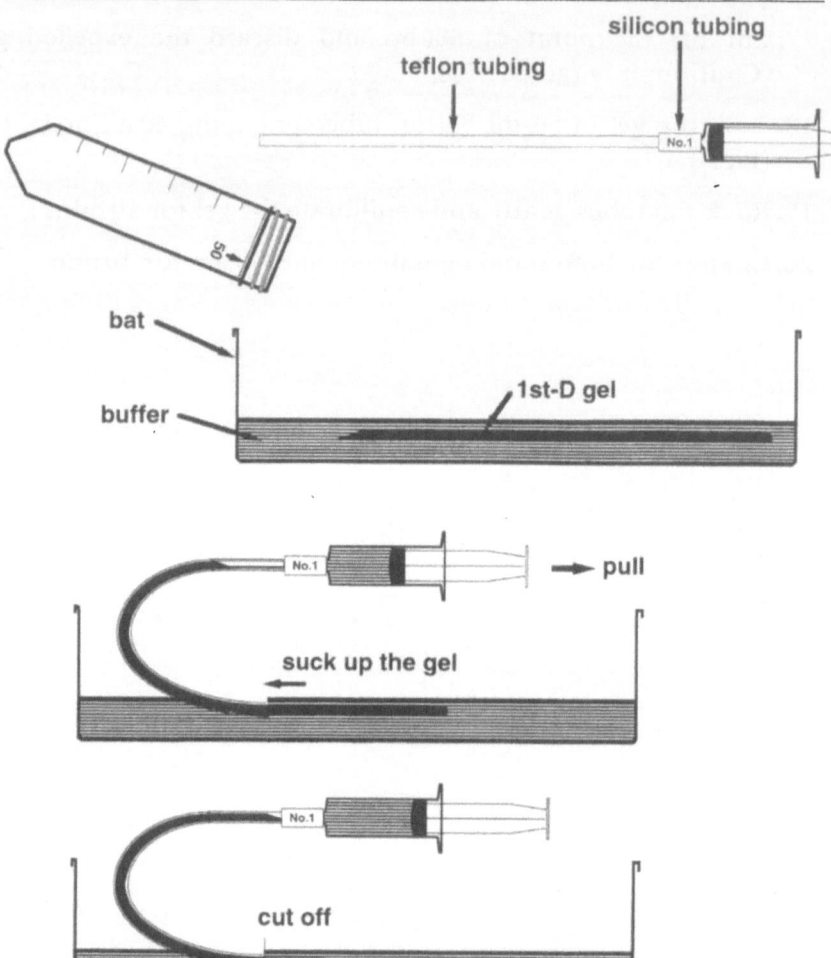

Fig. 3.5. Method for sucking up the 1-D gel noodle into digestion tube. Original method of 1st-D electrophoesis using Teflon tubing capillary was developed by Dr. Asakawa [1,2], and modified at RIKEN as shown in this figure

In Situ Restriction Enzyme Digestion (Fig. 3.5)

Enzyme digestion 21. Pour the equilibrated gel into a stainless tray containing 20 ml of 1× HB.

22. Gently suck up the gel noodle into the Teflon tubing.

23. Remove any traces of the buffer.

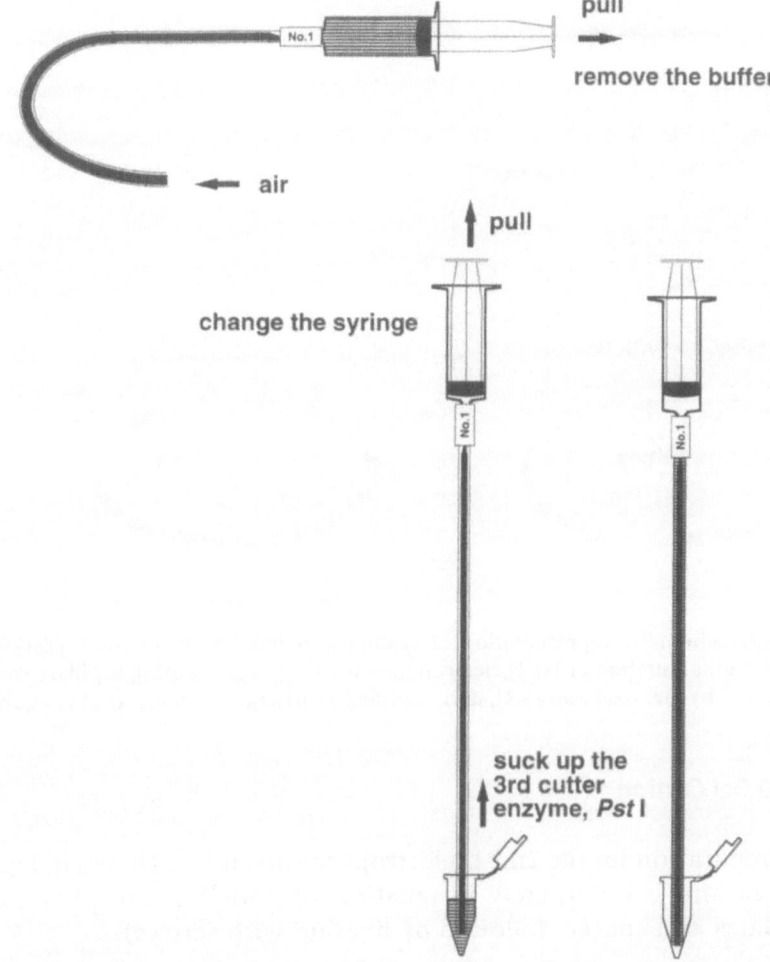

Fig. 3.6. Method for filling up the digestion tube with reaction mixture of restriction enzyme E_c. Original method of 1st-D electrophoresis using Teflon tubing capillary was developed by Dr. Asakawa [1,2], and modified at RIKEN as shown in this figure

24. Fill the tubing with 1500 µl of 1× HB containing 1500 units of *Pst*I and 0.01% BSA by sucking up slowly (Fig. 3.6).

25. Remove the syringe and loop the tubing by connecting one end to the other.

26. Put the looped tubing into a nylon bag (Fig. 3.7).

27. Seal the bag and incubate for 2 h at 37 °C (*do not exceed 2 h*).

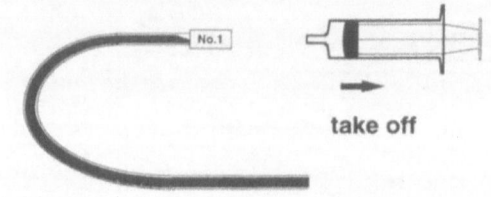

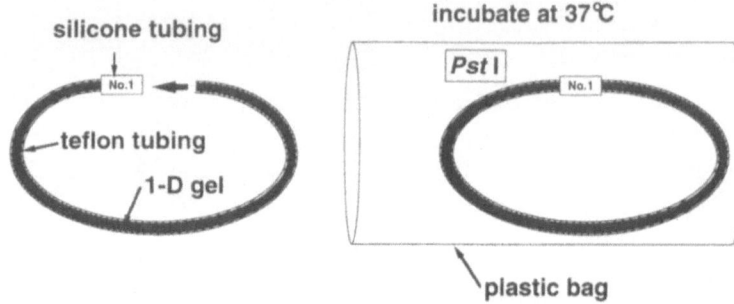

Fig. 3.7. Schematic representation of sealing and incubation of the digestion tube. Original method of 1st-D electrophoresis using Teflon tubing capillary was developed by Dr. Asakawa [1,2], and modified at RIKEN as shown in this figure

2nd-D Gel Casting

Gel casting 28. Preparation for the 2nd-D electrophoretic tank as shown in Fig. 3.2a and b (alternately installation of 2nd-D electrophoretic plates and spacers followed by fixation with screws).

29. Seal the side holes with Scotch 3M tape.

30. Mix the following agents for gel casting

Acrylamide	96.7 g
Methylenebisacrylamide	3.3 g
Ammonium persulfate	1.3 g
10× TBE	200 ml
DW	approx. 1600 ml

31. Stir.

32. Adjust the volume to 2000 ml with DW.

33. Add 800 µl TEMED just before use to solidify the gel.

34. Pour for 2D electrophoretic tank.

35. Drop 2-butanol on top of the gel.

36. Wait for about 2 h.

37. Peel away the Scotch 3M tape.

38. Wash out 2-butanol.

Running the 2nd-D Gel

39. Rinse the top of the gel with 1× TBE and wipe with Kimwipe for pretreatment of the 2nd-D gel.   **Gel running**

40. After 2 h digestion with *Pst*I expel the gel noodle from the tubing into a 50-ml Falcon tube containing 40 ml of 1× TBE.

41. Equilibrate for 10 min.

42. Discard the buffer and pour the gel noodle onto a plastic board.

43. Transfer the gel onto the top of 2nd-D acrylamide gel.

44. Connect agarose gel noodle and acrylamide gel with 2 ml of connecting gel (0.8% agarose in 1× TBE/0.05× dye solution) using a 5-ml syringe with 26- or 22-gauge needle.

45. Electrophoresis at 100 V for 2 h and 150 V for 24 h until the BPB reaches the bottom.

Gel Drying and Autoradiography

46. Dry the gel at 80 °C.   **Autoradio-graphy**

47. Autoradiograph using XAR 5 (Kodak) (3–14 days for long exposure).

3.3
Protocol for Combination 2 (*Pac*I-*Eco*RV-*Mbo*I)

3.3.1
Restriction Enzyme Digestion and Labeling

– ddATP[α]S (Toyobo)   **Materials**

– terminal transferase (Toyobo)

– [α-^{32}P] ddATP (5000 Ci/mmol) (Amersham, UK)

- *Pac*I (NEB)

- *Eco*RV (Takara)

- *Pst*I (Takara)

- *Sca*I (Takara)

Stock buffers
- 10× high buffer (HB)

- 10× medium buffer (MB)

- TE buffer

- 6× dye solution

- 10× K buffer (KB) (200 mM Tris-HCl pH 8.5, 100 mM MgCl$_2$, 1 M KCl, 10 mM DTT)

Protocol   Blocking

Blocking   1. Mix the following agents

Sample (0.5–1.5 µg/µl DNA)	100 µl
1 mM DTT	20 µl
1 M sodium cacodylate (pH 7.5)	20 µl
10 mM CoCl$_2$	20 µl
1 mM ddATPαS	2 µl
Terminal transferase (8 U/ml)	3 µl
Total	165 µl

2. Mix by pipetting.

3. Incubate for 30 min at 37 °C.

4. PCI extraction.

5. EtOH precipitation.

6. Add 50 µl of TE.

PacI Digestion

7. Mix the following agents

Sample	50 µl
10× MB	10 µl
0.1% BSA	10 µl
PacI (2 U/µl)	10 µl
DW	20 µl
Total	100 µl

8. Mix by pipetting

   90 µl

   10 µl
   Control: Use 10 µl of the
      mixture for cut check
   Add 1 µl of 1 µg/µl pNEB 193/
      ScaI cut
   Mix by pipetting

9. Incubate for 4 h at 37 °C

   Incubate for 4 h at 37 °C
   0.8% agarose electrophoresis
   (If DNA is completely cut, two
      bands are seen)

Labeling

10. Mix the following agents

Sample	10 µl
1 mM DTT	3 µl
1 M sodium cacodylate (pH 7.5)	3 µl
10 mM CoCl$_2$	3 µl
[α-^{32}P] ddATP (5000 Ci/mmol)	9 µl
Terminal transferase (8 U/µl)	2 µl
Total	30 µl

11. Mix by pipetting.

12. Incubate for 30 min at 37 °C.

13. PCI extraction.

14. EtOH precipitation.

15. Add 20 µl of TE.

*Eco*RV Digestion

Digestion   16. Mix the following agents

Sample	20 µl
HB	5 µl
*Eco*RV (20 U/µl)	5 µl
DW	20 µl
Total	50 µl

17. Mix by pipetting (50 µl)

48 µl

2 µl
Control: Use 2 µl of the
mixture for cut check
Add 8 µl of 0.1 µg/8 µl pBSII/
*Sca*I cut in HB
Mix by pipetting

18. Incubate for 1 h at 37 °C

Incubate for 1 h at 37 °C
0.8% agarose electrophoresis
(If DNA is completely cut, two
bands are seen)

19. Adjust the concentration of DNA to 0.2 µg/µl by adding 6× dye solution (final concentration: 1.5×) and TE

Two-dimensional separation with eletrophoresis and in situ restriction enzyme digestion is the same as combination 1 except for the reaction buffer for the enzyme. K buffer is appropriate for *Mbo*I.

References

1. Hayashizaki Y, Hirotsune S, Okazaki Y, Muramatsu M, Asakawa J (1995) Restriction landmark genomic scanning (RLGS), Molecular Biology and Biotechnology. VCH Weinheim, pp 813–817
2. Hayashizaki Y, Hirotsune S, Okazaki Y, Muramatsu M, Asakawa J. Restriction landmark genomic scanning (RLGS), Encyclopedia of Molecular Biology, vol 6. VCH, Weinheim (in press)
3. Okazaki Y, Okuizumi S, Sasaki N, Ohsumi T, Kuromitsu J, Kataoka H, Muramatsu M, Iwadate A, Hirota N, Kitajima M, Plass C, Chapman VM, Hayashizaki Y (1994) A genetic linkage map of the mouse using an expanded production system of restriction landmark genomic scanning (RLGS Ver. 1.8). Biochem Biophys Res Commun 205:1922–1929

Chapter 4

RLGS Spot Cloning

Sachihiko Watanabe and Yoshihide Hayashizaki

Contents

4.1
Breakthrough in the Development of RLGS Spot Cloning

In order to realize the potential of this technique, it is essential to have a direct and simple method for recovering individual loci of interest. A direct method for recovering landmark spots or sites is also essential for transferring the information to other laboratories and for identifying the gene functions that may be associated with these sites. The following points must be considered in the recovery of sequences associated with an RLGS spot.

First, the only tag for identifying a landmark is the restriction site sequence at the end of the target DNA fragment.

Second, the RLGS landmarks in genomic DNA that are identified as spots in the two-dimensional gel are surrounded by unlabeled DNA fragments which are 2000 times more abundant than the labeled sequence fragments. Thus, the direct cloning or amplification of landmark sequences will have a very high background of similar-sized DNA fragments [1,2].

Third, only a very limited amount of the spot sequence exists at the site of the labeled DNA because only 1.5 μg of whole genomic DNA is used for RLGS analysis, to keep spots defined and well separated. Thus, less than 0.75 attomoles (7.5×10^{-19}) of target DNA fragments are available for cloning or PCR amplification from the gel even if all the DNA molecules are recovered.

To overcome these problems, two methods have been developed. One is a restriction trapper-mediated method, in which genomic DNA fragments that contain the restriction landmark sites at their ends [3–5] are purified from restriction landmark-second cutter double digests. After this purification step, only the purified DNA fragments are subjected to RLGS, followed by the punching out of those spots of interest from the gel and cloning them, resulting in the efficient elimination of background DNA fragments.

The other is a PCR-mediated method in which the DNA fragments punched out from the RLGS gel are ligated to a PCR adapter and amplified, followed by the elimination of the primer adapter with polyacrylamide gel electrophoresis and plasmid cloning.

4.2
RLGS Spot Cloning by the Restriction Trapper-Mediated Method

4.2.1
Principle of the Restriction Trapper-Mediated Method

Figure 4.1 shows the structure of the restriction trapper. A hairpin-looped oligolinker is covalently linked to the surface of latex beads. For binding to latex beads, a ten-base cytosine stretch looping out from the hairpin formation is the donor of the amino residue. The beads are the carboxyl residue donor. This oligolinker carries the *Not*I site at its end. The principle of purification of DNA fragments with *Not*I (sites) at their end(s) is shown in Fig. 4.2. The genomic DNA digested with *Not*I and second cutter *Eco*RV are ligated to the restriction trapper. After the nonligated fragments have been washed out, the target fragments are cut out again with *Not*I and recovered. Even is contaminative DNA fragments are illegitimately ligated to a linker carrier, the target DNA fragments are strictly selected in the recutting step. The selectivity of the restriction trapper is based on the highly restricted selectivity of the restriction enzyme cleavage. Therefore, the target fragments can be concentrated very efficiently (more than 1×10^6-fold) with a minimum bias for the purification of DNA fragments less than 6 kb long [3].

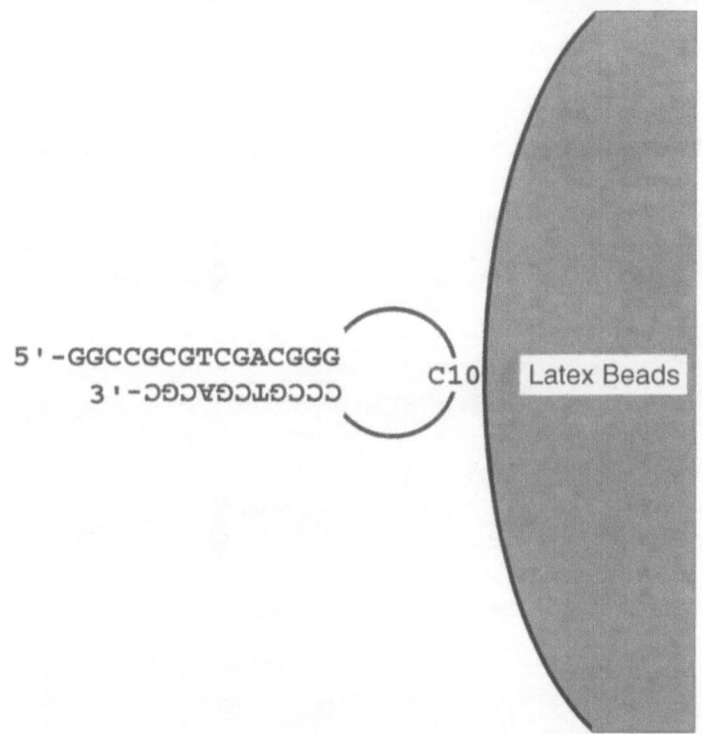

Fig. 4.1. Structure of the restriction trapper. A 36-bp oligonucleotide was covalently linked to the surface of latex beads. A ten-base cytosine stretch looping out from a hairpin formation is the donor of the amino residue for binding to latex beads (beads are carboxyl residue donor). This oligolinker carries the *Not*I site as its end and also has the *Sal*I site [8]. (By permission of VCH)

The method for RLGS spot target cloning is outlined in Fig. 4.3. Because the final genomic DNA fragments in the gel are products of the digestion of these three enzymes (*Not*I, *Eco*RV, and *Mbo*I), the fraction of recovered DNA from the punched-out gel contains more than 2000-fold unlabeled *Eco*RV-*Eco*RV, *Eco*RV-*Mbo*I, and *Mbo*I-*Mbo*I fragments, which produce many background clones. To eliminate such background fragments, the *Not*I restriction trapper was employed as shown in Fig. 4.2 [3,4]. Subsequently, the DNA fragments with *Not*I sites at their ends are subjected to RLGS, producing corresponding patterns. Then, the spots of interest should be punched out, and corresponding DNA fragments are recovered and subjected to direct cloning using plasmid vectors.

This restriction trapper-mediated method has three advantages. Firstly, selection of the *Not*I fragments reduced the

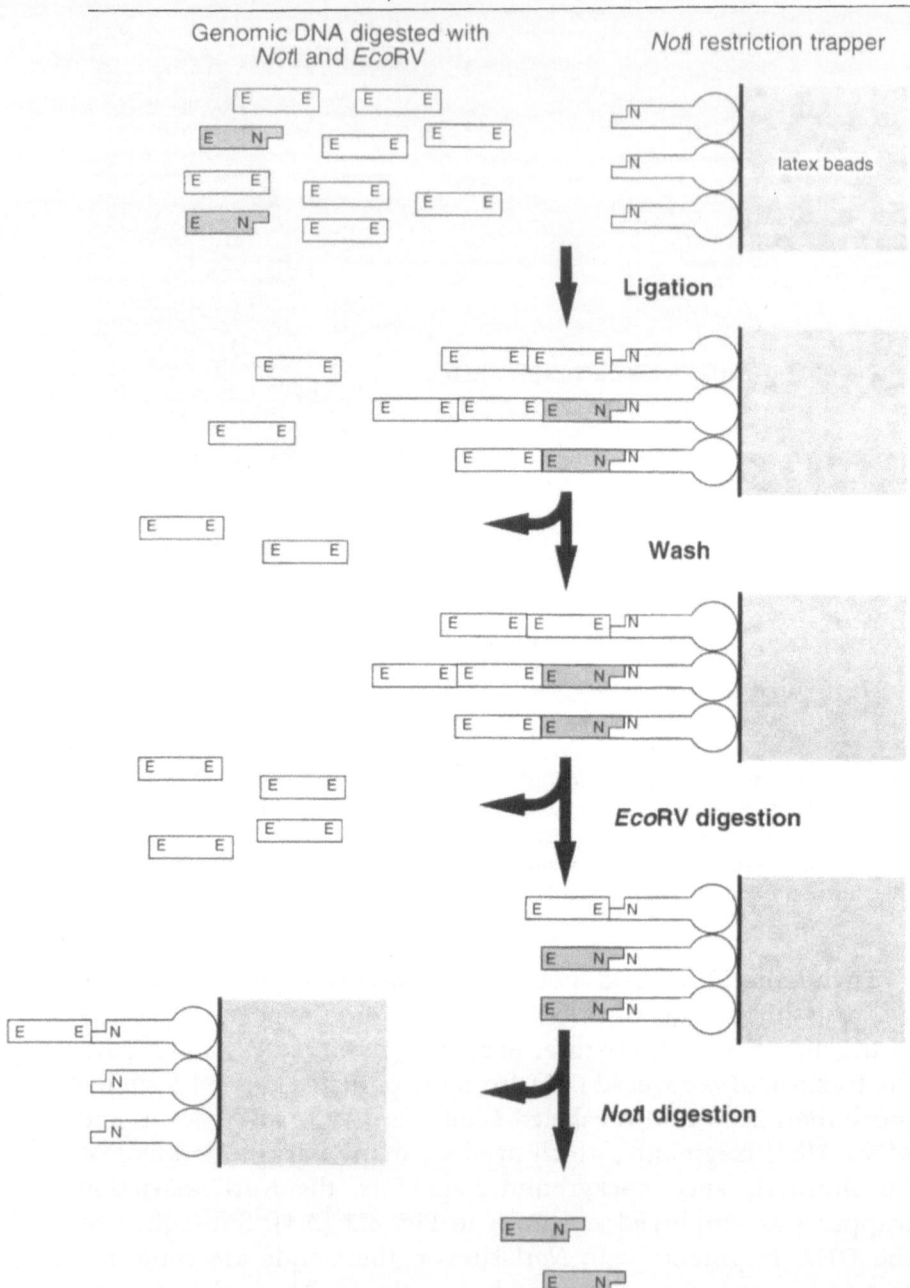

Fig. 4.2. Principle of the purification of a target DNA fragment with the restriction trapper [8]. (By permission of VCH)

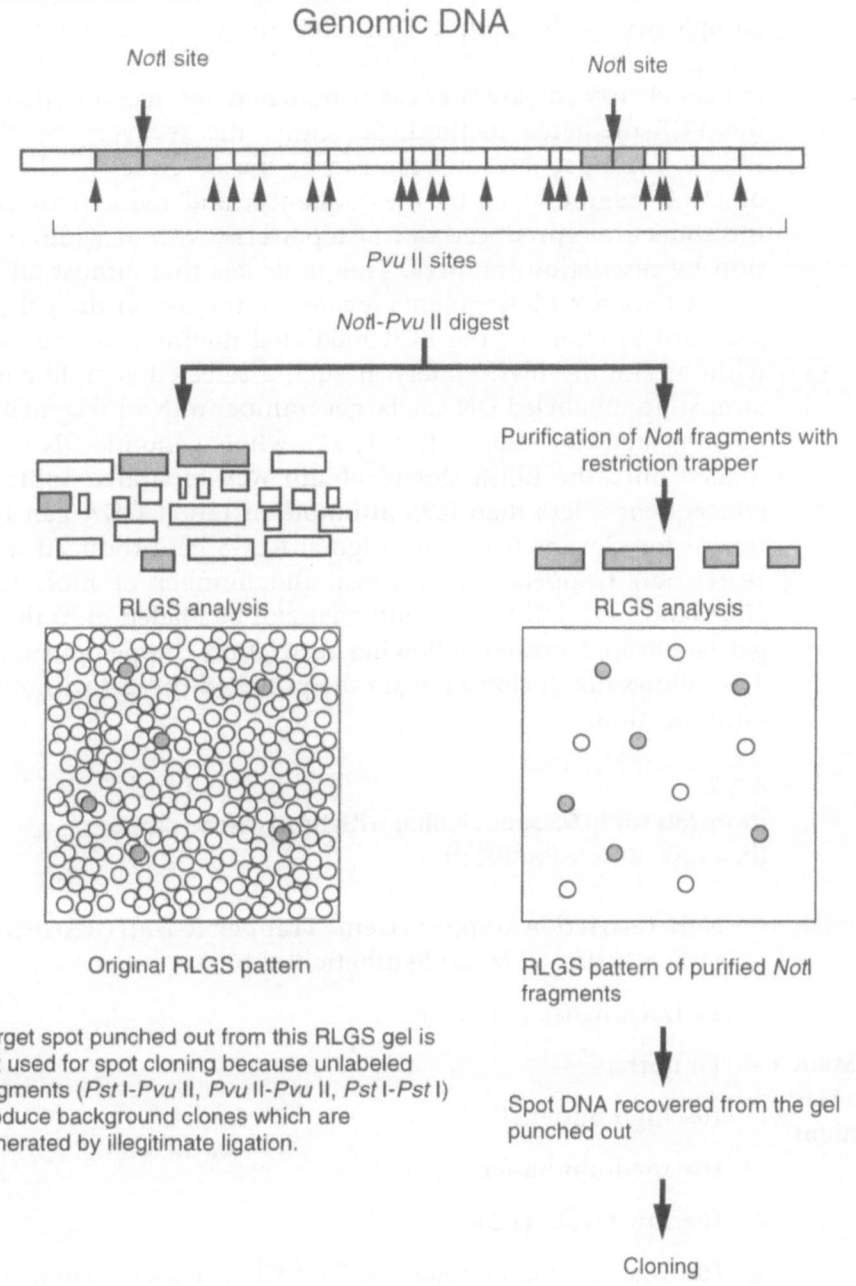

Fig. 4.3. Procedure of the RLGS spot target cloning method. *Not*I, *Eco*RV, and *Mbo*I were used as restriction enzymes E$_L$, E$_B$, and E$_C$, respectively. *Not*I-*Eco*RV fragments purified by the restriction trapper were used for RLGS analysis. After punching out a spot from the RLGS profile, the eluted DNA fragments were ligated with *Not*I-*Bam*HI digested pBlueScriptII. [8] (By permission of VCH)

complexity of the sample DNA, resulting in a reduction of the background on the RLGS profile, as shown in Fig. 4.3. The spots appear clearly and are very easily punched out in comparison with the PCR-mediated method. Secondly, the recovery by the restriction trapper does not tend to be biased either by the length of DNA fragments or by the sequence, and more than 90% of the spots are reproduced in RLGS patterns with or without selection by restriction trappers. This indicates that almost all spots, even GC-rich 2-kb fragments located at the top of the gel, which are hard to clone by the PCR-mediated method, can be isolated without cloning bias. Finally, if such a selected sample contains almost no unlabeled DNA, a larger number of *Not*I fragments can be subjected to RLGS. Only 1.5 µg of whole genomic DNA can be loaded onto the RLGS slot to obtain well-separated spots. As a consequence, less than 0.75 attomole of target DNA can be obtained for cloning from the original RLGS gel produced without restriction trappers. By contrast, the number of molecules of *Not*I landmark DNA fragments that can be loaded onto the same gel is 200-fold greater following restriction trapper purification. This allows direct cloning from the punched-out gel without PCR amplification.

4.2.2
Protocols for RLGS Spot Cloning with Restriction Trapper
(in a case of *Not*I-*Pvu*II-*Pst*I)

Materials – *Not*I restriction trapper (Gene Trapper R-*Not*I) (AGENE Research Institute, Japan Synthetic Rubber)

– T4 DNA ligase (Takara)

Stock – TE buffer
buffers
solutions – 10× high buffer (HB)

– 10× medium buffer (MB)

– 10× low buffer (LB)

– 10× ligation buffer [660 mM Tris-Cl (pH 7.4), 66 mM MgCl$_2$, 100 mM DTT]

– 6× dye solution [0.25% BPB, 0.25% XC in TE without glycerol]

– 2 M NaCl

- 5 M NaCl

- 100 mM ATP

- 50% polyethylene glycol 6000 (PEG)

- 3 M sodium acetate

- 1% linearly polymerized acrylamide

- Elution dye [7.5 M ammonium acetate with 1× dye solution]

- PCI (distilled phenol : chloroform : isoamylalcohol = 50 : 49 : 1, saturated with TE, contained 0.1% 8-hydroxyquinoline)

Wash the Restriction Trapper with TE

1. Pick up 140 µl (10% solution of restriction trapper).   **Buffer change**

2. Add 260 µl TE.

3. Centrifuge at 15 000 rpm for 1.5 min.

4. Discard the supernatant.

5. Resuspend in 280 µl TE (5% solution).

Restriction Enzyme Digestion (Ten Parallel Samples)

6. Mix the following agents   **Enzyme digestion**

Genomic DNA (100 µg)	x µl
10× HB	20 µl
0.1% BSA	20 µl
0.1% Triton X-100	20 µl
*Not*I (8 U/µl)	10 µl
DW	(130-x) µl
Total	200 µl × 10 tubes

7. Incubate for 2 h at 37 °C.

8. Add the following agents.

10x LB	20 µl
*Pvu*II (12 U/µl)	10 µl
DW	170 µl
Total	400 µl

9. Incubate for 2 h at 37 °C.

10. Extract with PCI.

11. Precipitate with EtOH.

12. Adjust the DNA concentration to 2 µg/ml.

Purification of *Not*I Fragments Using Restriction Trapper (Eight Parallel Samples)

Purification   13. Mix the following agents

Samples (*Not*I-*Pvu*II digested)	50 µl
*Not*I restriction trapper (5% solution)	20 µl
10× ligation buffer	15 µl
2 M NaCl	15 µl
100 mM ATP	2 µl
50% PEG	30 µl
T4 Ligase (350 U/µl)	4 µl
DW	14 µl
Total	150 µl × 8 tubes

Mix well by pipetting.

14. Add 30 µl of 50% PEG 6000 in each tube and mix well by pipetting.

15. Incubate for more than 2 h at 18 °C, mix every 30 min.

16. Add the following agents

10× LB	45 µl
*Pvu*II (12 U/µl)	10 µl
DW	395 µl
Total	600 µl

17. Incubate for 1 h at 37 °C.

18. Incubate for 15 min at 65 °C.

19. Add 60 µl 0.1% Triton X-100.

20. Centrifuge at 15 000 rpm for 5 min.

21. Discard the supernatant.

22. Resuspend in 50 µl 1× MB.

23. Collect the eight tubes and divide them into two tubes.

24. Add 20 µl *Pvu*II.

25. Incubate for 1 h at 37 °C.

26. Add 200 µl DW and 40 µl 0.1% Triton-X100.

27. Centrifuge at 15 000 rpm for 5 min.

28. Discard the supernatant.

29. Resuspend in 200 µl 1× HB and 20 µl 0.1% Triton X-100.

30. Collect the two tubes into one new tube.

31. Centrifuge at 15 000 rpm for 5 min.

32. Discard the supernatant.

33. Add the following agents

10× HB	10 µl
DW	60 µl
0.1% BSA	10 µl
0.1% Triton X-100	10 µl

Mix well by pipetting.

34. Add 10 µl of *Not*I (10 U/µl).

35. Incubate for more than 2 h at 37 °C, mix every 30 min.

36. Centrifuge at 15 000 rpm for 5 min.

37. Pick up supernatant.

38. Resuspend precipitate in 100 µl of TE and recover supernatant by centrifugation at 15 000 rpm for 5 min.

39. Mix the supernatants together.

40. Extract with PCI.

41. Precipitate with EtOH in 2 µl 1% linearly polymerized acrylamide.

42. Dissolve in 16 µl TE.
    Control: Pick up 1 µl and check concentration of DNA.

Labeling

Labeling   43. Mix the following agents

Samples	3 µl
10× MB	1 µl
[α-^{32}P]dGTP	1 µl
[α-^{32}P]dCTP	1 µl
Sequenase Ver. 2.0	1 µl
DW	3 µl
Total	10 µl

44. Incubate for 30 min at 37 °C.

45. Incubate for 30 min at 65 °C.

46. Add 10 µl 6× dye and 8 µl TE.

    Mix well.

47. Add nonlabeled DNA 12 µl.

1st-D and 2nd-D Electrophoresis in RLGS Method (see Chap. 3)

Electrophoresis   48. Apply 10 µl of labeled samples on 1st-D slot.

49. *Pst*I is used for 3rd digestion; carry out 2nd-D electrophoresis.

50. Dry the gel at 65 °C.

51. Staple the dried gel with Kodak XAR 5 in the dark room.

52. Autoradiograph at −80 °C overnight.

53. Remove the staples and develop the film.

54. Adjust the film to the gel and restaple.

55. Punch out the target spot.

Recovering the DNA Fragment from the Punched-Out Gel by Electroelution

Electroelution   56. Soak the gel in 20 µl TE.

57. Pour 100 µl elution dye into the V-shaped slot.

58. Electroelute at 200 V for 20 min.

59. Pick up 300 μl of elution dye.

60. Extract with PCI.

61. Mix the following agents

3 M Na acetate (pH 5.2)	30 μl
1% linearly polymerized acrylamide	2 μl
Cold EtOH	700 μl

(1% linearly polymerized acrylamide is used as carrier).

62. Incubate at −80 °C for 15 min.

63. Centrifuge at 15 000 rpm for 30 min.

64. Resuspend in 3.5 μl TE.

Preparation of *NotI-PstI* Vector

65. Digest the pBSII with *Not*I and *Pst*I.

66. Electrophoresis; recover the vector fragment (dephospholyla-tion was not carried out).

Preparation of vector

Ligation

67. Mix the following agents

Spot DNA solution	3.5 μl
*Not*I-*Pst*I vector (50 pg/μl)	2 μl
10× ligation buffer	1 μl
100 mM ATP	0.1 μl
5 M NaCl	0.4 μl
T4 DNA ligase (350 U/μl)	1 μl
50% PEG 6000	2 μl
Total	10 μl

Ligation

68. Ligate overnight at 18 °C.

69. Transform.

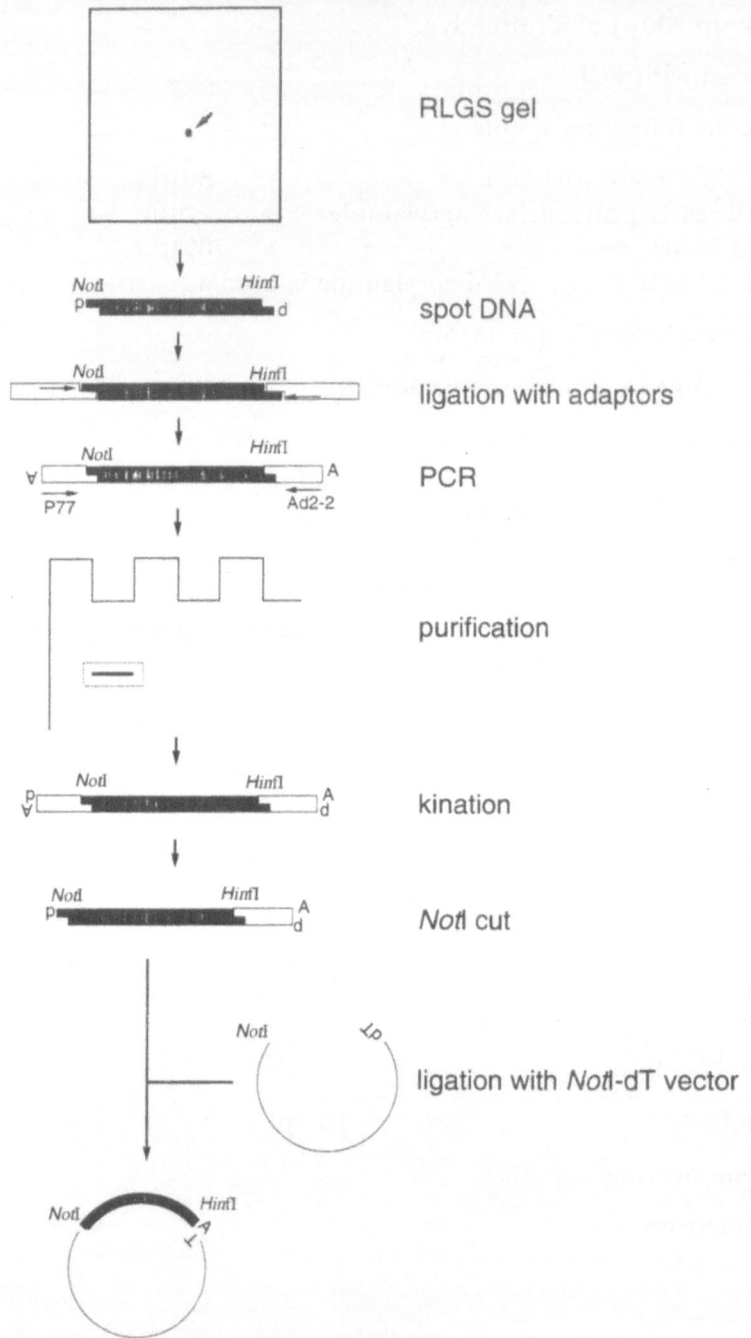

RLGS gel

spot DNA

ligation with adaptors

PCR

purification

kination

*Not*I cut

ligation with *Not*I-dT vector

Fig. 4.4. Strategy of a PCR-mediated method for cloning spot DNA

4.3
RLGS Spot Cloning by the PCR-Mediated Method

4.3.1
Principle of the PCR-Mediated Method

Almost all spots on RLGS profile can be cloned by the restriction trapper-mediated method; however, this method requires a large amount of DNA as starting material. The DNA starting material is sometimes limited, depending on the aim of experiments. In that case, the PCR-mediated method described in this chapter is very useful.

This strategy employs PCR amplification of adaptor-ligated spot DNA fragments followed by ligation with a *Not*I-dT vector consisting of a linearized and dephosphorylated pBluescriptII (Stratagene) with both a *Not*I terminus and 3′-dT protruding terminus (Fig. 4.4). The target spot DNA is punched out without excluding similar-sized DNA fragments, colocated on the RLGS gel. This is because such similar-sized DNA fragments are expected to work both as DNA carriers to prevent loss of the target spot DNA and as colocalized markers for this DNA. Furthermore, the amplified target spot DNA is selectively cloned from similar-sized DNA by ligation with the *Not*I-dT vector, since similar-sized DNA does not possess a *Not*I site at its end. This method should be applicable to any spot DNA which is derived from RLGS by using different sets of restriction enzymes for analysis, simply by preparing adaptors, PCR primers, and vectors for subcloning.

4.3.2
Protocol for the PCR-Mediated Method

- *E. coli* DNA ligase (Takara)   **Materials**

- Taq DNA polymerase (Takara)

- T4 polynucleotide kinase (Takara)

- T4 DNA ligase (Takara)

- Sephacryl S-300 (Pharmacia)

- NACS column (BRL)

- *Not*I-dT vector (linearized dephosphorylated pBSII with both a *Not*I terminus and 3′-dT protruding terminus)

- *Not*I adaptor (consisting of 5′-ACGCCAGGGTTTT-CCCAGTCACGACGC-3′ and 5′-pGGCCGCGTCGTGACTGGG-AAAACCCTGGCGT-3′)

- *Hin*fI adaptor (consisting of 5′-pANTCTGTACTGCACCAGC-AAATCC-3′ and 5′-GGATTTGCTGGTGCAGTACAG-3′)

- P77 (PCR primer), 5′-AGGGTTTTCCCAGTCACGACGCGG-3′

- Ad2-2 (PCR primer), 5′-TTGCTGGTGCAGTACAGANTC-3′

- P8 primer (Toyobo)

Stock buffers and solutions

- TAE buffer

- TE buffer

- $T_{10}E_{0.1}$ buffer– TE buffer

- 10× high buffer (HB)

- 10× medium buffer (MB)

- 10× low buffer (MB)

- 10× ligation buffer (for T4 DNA ligase) [500 mM Tris-HCl (pH 7.4), 100 mM MgCl$_2$, 10 mM DTT]

- 10× ligation buffer (for *E. coli* DNA ligase) [300 mM Tris-HCl (pH 8), 40 mM MgCl$_2$, 12 mM EDTA, 100 mM (NH$_4$)$_2$SO$_4$, 1 mM NAD, 0.05% BSA]

- 10× PCR buffer [100 mM Tris-HCl (pH 8.3), 500 mM KCl, 15 mM MgCl$_2$]

- dNTP mixture (each 2.5 mM)

- 10× kination buffer [500 mM Tris-HCl, 100 mM MgCl$_2$, 50 mM DTT]

- 100 mM ATP

- 50% polyethylene glycol 6000 (PEG)

- 3 M sodium acetate

- 1% linearly polymerized acrylamide

Recovering DNA from the Punched-Out Target Spot

1. Soak the gel in less than 100 μl TE.
2. Electroelute.
3. Extract with PCI.
4. Precipitate with cold EtOH by adding 1 μl of 1% linearly polymerized acrylamide.
5. Suspend in 3 μl TE.

Recovering of DNA

Ligation with Adaptors Using *E. coli* DNA Ligase

6. Mix the following agents

Ligation

DNA solution	1 μl
10× ligation buffer	0.5 μl
Adaptor for *Not*I (10 pmol/μl)	0.5 μl
Adaptor for *Hin*fI (10 pmol/μl)	0.5 μl
E. coli DNA ligase	1 μl
DW	1.5 μl
Total	5 μl

7. Ligate overnight at 16°C.
8. Add 35 μl of $T_{10}E_{0.1}$ (40 μl total).
9. Pass through Sephacryl S-300 spin column (exclusion of adaptors).

PCR Amplification Using Primers, Ad2-2 and P77

10. Mix the following agents

PCR amplification

Templete (ligation mixture)	37.5 μl
dNTP Mixture	5 μl
10× PCR buffer	5 μl
P77 primer (10 pmol/μl)	1 μl
Ad2-2 primer (10 pmol/μl)	1 μl
Taq DNA polymerase	0.5 μl
Total	50 μl

11. 30 cycles of 94 °C for 1 min, 60 °C for 1.5 min and 72 °C for 2 min.

Purification of PCR Product

Purification   12. Electrophoresis.

13. Electroelute in final volume about 300 µl and use 1/40–1/100 of electroeluted PCR product for PCR amplification.

14. Pass through NACS column (final volume about 500 µl).

15. Precipitate with EtOH.

16. Suspend in 10 µl TE.

Kination

Kination   17. Mix the following agents

Purified PCR product	10 µl
10× kination buffer	5 µl
100 mM ATP	1 µl
DW	30 µl
T4 DNA polynucleotide kinase (10 U/µl)	4 µl
Total	50 µl

18. Incubate for 60 min at 37 °C.

19. Extract with PCI.

20. Precipitate with EtOH.

21. Suspend in 5 µl TE.

*Not*I Digestion

Digestion   22. Mix the following agents

DNA solution	5 µl
10× HB	1 µl
DW	3 µl
*Not*I (10 U/µl)	1 µl
Total	10 µl

23. Incubate for 60 min at 37 °C.

24. Extract with PCI.

25. Precipitate with EtOH.

26. Suspend in 3.5 μl TE.

Ligation with NotI-dT Vector Using T4 DNA Ligase

27. Mix the following agents   **Ligation**

Spot DNA solution	3.5 μl
NotI-dT vector (50 ng/μl)	2 μl
10× ligation buffer	1 μl
100 mM ATP	0.1 μl
T4 DNA ligase (350 U/μl)	1 μl
50% PEG 6000	1 μl
DW	1.4 μl
Total	10 μl

28. Ligate overnight at 16 °C.

29. Transform into E. coli HB 101.

Direct PCR Amplificiation from Picked Colonies, Using Primers, Ad2-2 and P8

30. Mix the following agents   **PCR**
                                                                amplification

dNTP Mixture	5 μl
10× PCR buffer	5 μl
P8 primer (10 pmol/μl)	1 μl
Ad2-2 primer (10 pmol/μl)	1 μl
Taq DNA polymerase	0.5 μl
DW	37.5 μl
Total	50 μl

31. Carry out PCR: 30 cycles at 94 °C for 1 min, 60 °C for 1.5 min, and 72 °C for 2 min.

Selection of PCR Product

32. Discriminate from similarly sized fragment using fingerprint and/or Southern hybridization and obtain DNA fragment of the target spot.

4.3.3
Practical Application of the PCR-Mediated Method

We have shown the results for cloning DNA fragments derived from four target spots (designated 15, 29, 91, and 106), whose intensities changed developmentally due to DNA methylation in the telencephalon of C3H/HeN mice [6,7]. After PCR amplification, products of expected sizes were observed on polyacrylamide gel in the lanes for spots 15, 91, and 106 (Fig. 4.5A). The

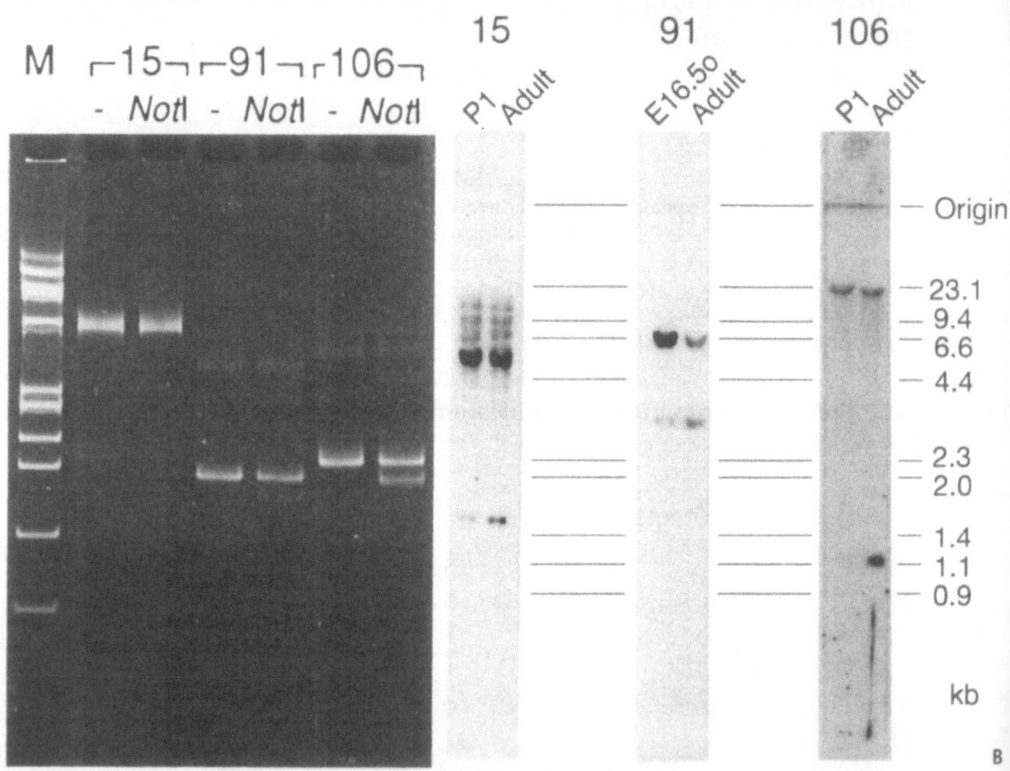

Fig. 4.5. Results of PCR-mediated spot cloning. **A** A typical electrophoretic pattern of PCR-amplified DNA fragments derived from three target spots. After second PCR, 5 or 10 μl of reaction was directly subjected to 8% polyacrylamide gel electrophoresis with 500 ng of φX174 *Hae*III DNA marker (lane *M*). **B** Genomic Southern blotting analysis for selection of the clones corresponding to the target spots. Five μg of *Bam*HI-*Not*I-digested genomic DNA was electrophoresed through 0.7% agarose gel, alkali-denatured, and then transferred onto Hybond-N nylon membrane (Amersham). Insert DNAs derived from each spot were labeled as probes, and hybridization was performed by the conventional method. Only the autoradiogram obtained by positive probes is shown. From H. Suzuki *et al.* [6], by permission of Oxford University Press

products were subcloned and utilized as probes for genomic Southern blot analysis. Because we observed no spots with changed intensities due to DNA methylation in developmental stages close to each target spot on the RLGS gel, positive clones were determined by

(1) the appearance of two hybridization bands due to partial digestion with the methylation-sensitive enzyme NotI,

(2) the consistency in size in the lower bands with the estimated size from the RLGS gel, and

(3) the alteration of the intensity of the lower band through development in C3H/HeN mice as observed on RLGS [6,7].

As shown in Fig. 4.5B, one major clone for each spot revealed the hybridization pattern we expected. Weak lower bands were observed in the lanes of postnatal day 1 (P1) for spot 15, and the outer layer of embryonal day 16.5 (E16.50) for spot 91, although neither target spot was detected on RLGS [6,7]. However, this may reflect the higher sensitivity of the Southern blotting as compared to RLGS. At present, 13 out of 20 (65%) target spots have been successfully cloned using our strategy. In spite of the difficulty in cloning target spot DNA of more than 1000 bp, such as spot 29, our method may be highly useful when the total DNA sample prepared for cloning is limited, since only a few micrograms of total DNA is sufficient for spot cloning. Moreover, the method seems to be more applicable for cloning less numerous spot DNA, for example, methylation affected ones, because it is based on PCR amplification of individual target spots.

References

1. Hatada I, Hayashizaki Y, Hirotsune S, Komatsubara K, Mukai T (1991) A genomic scanning method for higher organisms using restriction sites as landmarks on the genome. Proc Natl Acad Sci USA 88:9523–9527
2. Hayashizaki Y, Hirotsune S, Okazaki Y, Hatada I, Shibata H, Kawai J, Hirose K, Watanabe S, Fushiki S, Wada S, Sugimoto T, Kobayakawa K, Kawara T, Katsuki M, Sibuya T, Mukai T (1993) Restriction landmark genomic scanning and its various applications. Electrophoresis 14:251–258
3. Hayashizaki Y, Hirotsune S, Hatada I, Tamatsukuri S, Miyamoto C, Furuichi Y, Mukai T (1992) A new method for constructing NotI linking and boundary library using restriction trapper. Genomics 14:733–739

4. Hirotsune S, Shibata H, Okazaki Y, Sugino H, Imoto H, Sasaki N, Hirose K, Okuizumi H, Muramatsu M, Plass C, Chapman VM, Miyamoto C, Tamatsukuri S, Furuichi Y, Hayashizaki Y (1993) Molecular cloning of polymorphic markers on RLGS gel using the spot target cloning method. Biochem Biophys Res Commun 194:1406–1412

5. Okuizumi H, Okazaki Y, Sasaki N, Muramatsu M, Nakashima K, Fan K, Tano H, Ohba K, Hayashizaki Y (1994) Application of the RLGS method to large-size genomes using a restriction trapper. DNA Res 1:99–102

6. Suzuki H, Kawai J, Taga C, Ozawa N, Watanabe S (1994) A PCR-mediated method for cloning spot DNA on restriction landmark genomic scanning (RLGS). DNA Res 1:245–250

7. Kawai J, Hirotsune S, Hirose K, Fushiki S, Watanabe S, Hayashizaki Y (1993) Methylation profiles of genomic DNA of mouse developmental brain detected by restriction landmark genomic scanning (RLGS) method. Nucleic Acids Res 21:5604–5608

8. Ohsumi T, Okazaki Y, Shibata H, Hirotsune S, Muramatsu M, Suzuki H, Taga C, Watanabe S, Hayashizaki Y (1995) RLGS spot cloning method, Electrophoresis 16:203–209

9. Hayashizaki Y, Hirotsune S, Okazaki Y, Muramatsu M, Asakawa J (1995) Restriction landmark genomic scanning (RLGS). In: Meyers RA (ed) Molecular biology and biotechnology: a comprehensive desk reference. VCH, Weinheim, pp 813–817

Chapter 5

RLGS Spot Mapping Method

Hisato Okuizumi, Yasushi Okazaki,
and Yoshihide Hayashizaki

Contents

5.1
Introduction

Recent progress in molecular genetics has been expedited by the development of DNA landmarks. The DNA landmarks visualized by Southern blot and PCR enabled us to construct high-density genome map which is essential for position-dependent identification of the gene responsible for a certain phenotype. This approach is so-called positional cloning. In the medical field, the information of the genes identified by position-dependent cloning (including positional candidate approach) is very useful for diagnosis, prevention, gene therapy, and drug therapy. This approach is the only known systematic way to connect any phenotypes to the transcripts, based on genetic analyses and molecular cloning technology. To facilitate the identification of genes using the positional cloning approach in higher organisms, high-density genetic maps which require a multiplex genome scanning method are essential. This is especially important in the case of orphan genomes which have important mutants but no dense map, so that we should employ a method that can rapidly scan a large number of loci.

In this section, we focus the RLGS spot mapping method with the mouse, C57BL/6 (B6 or B), DBA/2 (D2 or D), and *Mus spretus* (Sp or S) as one of the many applications of RLGS. The genetic analysis of RLGS loci demonstrates the distribution of these landmarks across the genome. The analysis of these sites employs the same approaches that would be used for RFLP analysis or, more specifically, the analysis of multigene families such as endogenous retroviruses, including intercisternal A particles [1], C-type retroviruses [2], B-type endogenous mammary tumor viruses, or Mtv sequences [3,4]. In these particular cases, strain differences in restriction patterns are able to be identified by hybridization with genetic probes for a viral subfamily. Subsequently, these strain-specific restriction fragments are followed in the backcross progeny or in recombinant inbred (RI) strain analyses. In numerous instances, the occurrence of a virus-like element in one strain may or may not have an allelic counterpart in another strain at the identical genomic site. Therefore, the analysis is dependent on the ability to follow the strain-specific locus as a dominant, segregating phenotype. Yet other multilocus markers also behave in comparable ways, such as the random amplification of polymorphic DNA (RAPDs) that has been characterized in the rat [5], the mouse [6–8], the zebra fish [9], etc. Additionally, to identify series of markers across the genome [10], motif sequence-tagged PCR products have been employed.

RLGS loci share the same characteristics in the primary genetic analysis as the retroviral subfamilies. There exist differences, however, in several important and significant features:

(1) When GC-rich enzymes are employed as landmark enzymes, RLGS landmarks are rare-site cleavages of genomic DNA that seem to be very closely related to CpG islands situated throughout the genome. Aside from the common sequence identification of the restriction cleavage site and the characteristic CpG-rich nature of these particular sequences, there is little or no sequence similarity in the labeled landmark fragments between different landmarks. Fundamentally, each RLGS locus typifies a unique and possibly functional gene.

(2) The sequences are common in the gene pool, even to the cross-species identification of landmark loci. Hence, boundary clones for an RLGS locus may serve as valuable reagents in the identification of the homologous locus in comparative analyses

[11]. Although strains or species may differ in the frequency of the landmark cleavage sequences, the surrounding sequences should remain highly conserved. The landmark cleavage site may be present in both parental strains of a cross, but the second or third restriction cleavage may recognize different flanking sequences. Additionally, it is possible that there are

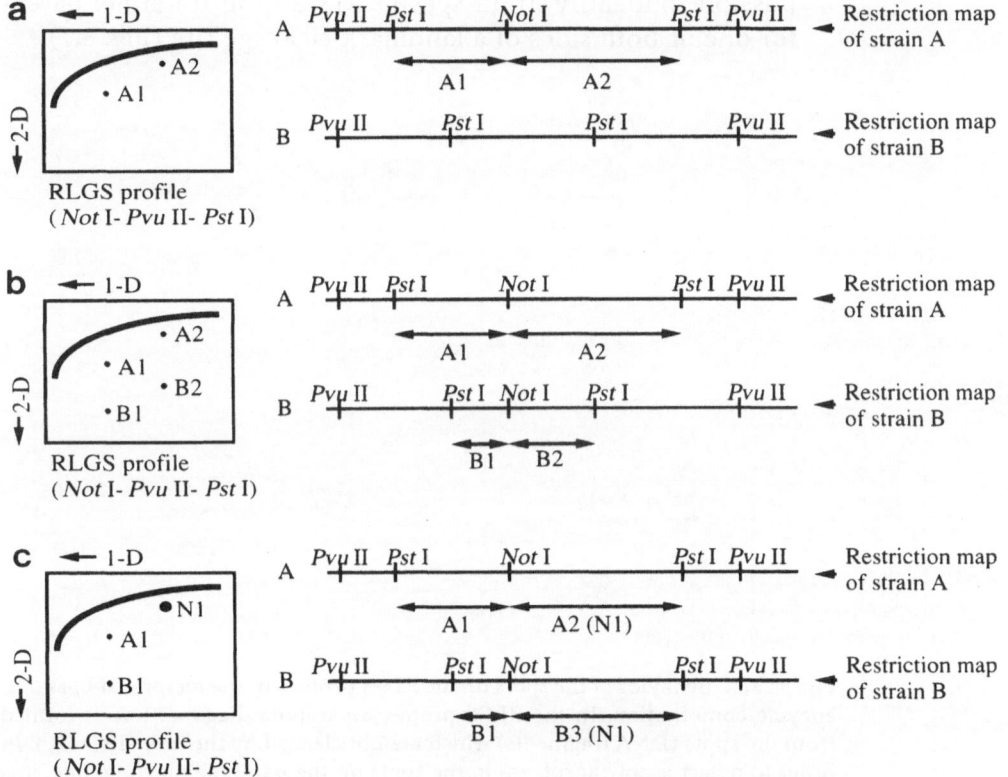

Fig. 5.1a–c. Behavior of the spot-polymorphism on the RLGS profile. In this figure, *Not*I-*Pvu*II-*Pst*I is employed as the enzyme combination of E_L-E_B-E_C in the RLGS profile of F_1 which is from the cross between A and B parental strains. **a** When *Not*I site is in strain A but not in strain B, two spots can be detected (Both A1 and A2 originated from strain A) in the RLGS profile. The DNA fragments of the spots correspond to the *Not*I-*Pst*I fragments as shown under the restriction map of strain A. **b** In the case of both strains A and B having a *Not*I site but different pairs of *Pst*I sites, four spots can be detected (A1 and A2 are the same spots as in **a**, B1 and B2 originate from strain B) in the profile. **c** If one side of the *Pst*I site of strain B is the same as A, the DNA fragments of A2 and B3 are represented as a nonpolymorphic spot, N1

insertions of repeat elements or deletions of sequences within the regions between the primary landmark site and either the second or third cleavage. Hence, variation in RLGS loci is equivalent to other restriction fragment length polymorphisms in a population.

(3) The use of primary cleavage enzymes that leave reciprocal overlapping ends such as *Not*I, that can be equally endlabeled, enhances the ability to identify genetic variation between strains for a given site. In these particular cases, it may be possible to identify strain-specific RLGS spots for either parent for one or both sides of a landmark cleavage site (Fig. 5.1).

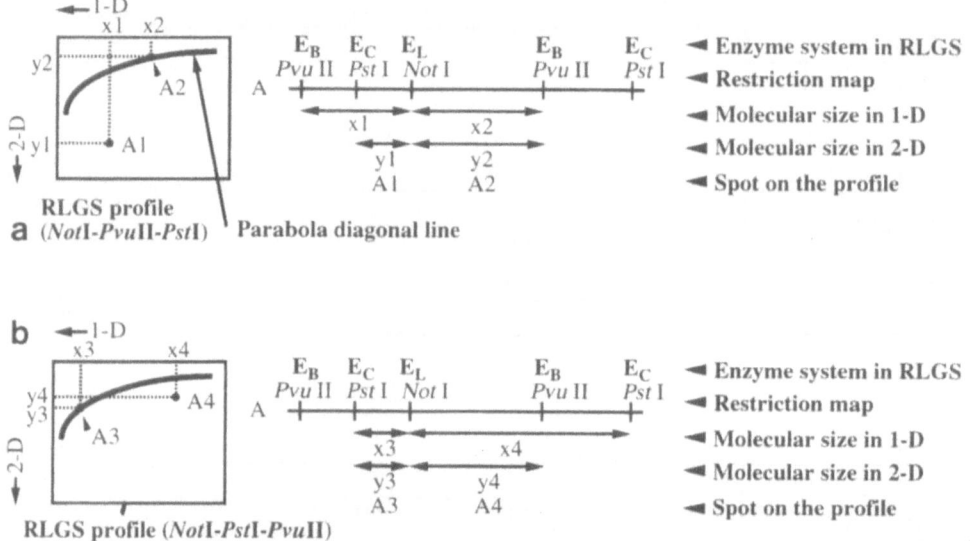

Fig. 5.2a,b. Behavior of the spots on the RLGS profiles in the reciprocal E_B and E_C enzyme combination. In the RLGS profile, a parabola diagonal line is formed from the spots (DNA fragments) which are not cleaved by the 3rd cutter (E_C). In order to detect a polymorphism in the spots on the parabola diagonal line, it is very advantageous to use a reciprocal enzyme combination for E_B and E_C such as *Not*I-*Pvu*II-*Pst*I and *Not*I-*Pst*I-*Pvu*II, which are shown in this figure as an example. The order of relative localization of enzyme sites is classified into two categories, as shown in this figure. **a** When the enzyme combination of *Not*I-*Pvu*II-*Pst*I is employed, the final DNA fragments from the locus represented by the spot *A1* are electrophoresed under the parabola diagonal line. In the case of *A2*, the DNA fragment of the spot A2 after the E_C cleavage procedure is the same as the one after the 2nd digestion (E_B). Consequently, the spot A2 is on the parabola diagonal line. **b** In the convert enzyme combination, *Not*I-*Pst*I-*Pvu*II, spot A4 corresponds to A2 in the profile a, represents a spot under the parabola diagonal line. On the other hand, A3 (corresponding to the spot A1) is on the parabola diagonal line

We can directly estimate the number of possible loci for a given restriction cleavage site such as *Not*I [12]. In the case of the laboratory mouse, there are an estimated 2380 *Not*I sites, and the estimated number of endlabeled landmarks is therefore 4760. If we can visualize approximately 2000 landmark spots on a single two-dimensional gel, it is estimated that there is a 66.4% ($1-(2760/4760)^2$) chance that one of the total *Not*I landmarks will, in fact, be sampled in any single analysis. Utilizing this number, it can be calculated that it will take approximately three different 2nd and 3rd restriction enzyme combinations to identify 95% of the *Not*I landmarks for a strain. In this case, it is very advantageous to use reciprocal enzyme combination for E_B and E_C, such as *Not*I-*Pvu*II-*Pst*I and *Not*I-*Pst*I-*Pvu*II (Fig. 5.2).

The genetic analysis of these landmarks will be dependent upon the relative variability between the parental strains employed for a given cross. We have previously estimated that the inbred strains DBA/2 (D2) and C57BL/6 (B6) have 13% variation in RLGS landmarks. Additional comparative studies have shown 26 and 31% variation between *M. molossinus* (MSM) and B6, and *M. musculus* (PWK) and B6, respectively [13]. More recently, variation between B6 and *M. spretus* has been analyzed. Results show that these mice differ for more than 50% of the RLGS loci [14]. If we combine the estimate of the variation between strains with the estimate of the number of enzyme combinations that must be used to analyze 95% of the landmarks, it is possible to determine the potential number of *Not*I landmarks that can be genetically identified and analyzed.

The genetic analysis of landmarks can be accomplished using interspecific backcrosses and RI strain analyses. The use of RI strains, the BXD series for example, is particularly valuable since it allows us the opportunity to simultaneously analyze both parental strains as strain distribution patterns (SDPs) and to compare these with the cumulative genetic map. In addition, we have also analyzed the segregation of B6-specific loci in a backcross of (B6 × *M. spretus*) F_1 × *M. spretus* (BSS) [14]. In principle, we should be able to identify *M. spretus*-specific RLGS loci in a comparison of the F_1 hybrid profile with B6 and *M. spretus*. The segregation of these *M. spretus*-specific loci should appear as quantitative differences in spot labeling in backcross progeny with homozygous *M. spretus*-specific genotypes as full intensity of labeling (double copy) and heterozygous progeny showing half intensity of labeling (single copy) [15].

Here, we summarize the progress made in the genetic analysis of RLGS loci and the progress made toward assigning these markers across the genome to specific chromosomal addresses in the mouse genome. The initial work has employed primarily RLGS methods in the genetic analysis. It is also possible, however, to expand this work and to use boundary clones for specific loci to confirm the spot clone assignment with standard Southern analysis methodology. Moreover, it is feasible to extend the use of these clones to either FISH analysis on human chromosomes or to somatic genetic analysis of rodent-human hybrid cell panels.

5.2
Interspecific Congenic Strain Analysis

We have analyzed a congenic strain of mouse C57BL/6(B6)-*Gus*^{s}. The B6-*Gus*^{s} congenic strain is genetically identical to B6 with the exception of a relatively small region surrounding the *Gus* gene. The B6-*Gus*^{s}, which has the *Gus* region from *M. spretus* (Sp), was produced by nine generations of selection and backcrossing. The congenic B6 strain carries a 4.2-cM region of *M. spretus* chromosome surrounding the *Gus* locus on chromosome 5.

The RLGS profiles of B6-*Gus*^{s}, *M. spretus*, and B6 were compared to identify B6-specific spots missing from the congenic strain and Sp-specific spots that were present. Figure 5.3 shows the schematic figure of the congenic strain of *Gus*. About 1000 to 2000 landmarks

Table 5.1. RLGS loci that are mapped to the *GusS* region as either C57BL/6 (B)-specific spots lost or *M. spretus* (S)-specific spots gained [16]

Enzyme combination	No. of screened spots	Candidate spots			Mapped spots		
		B	S	Total	B	S	Total
*Not*I-*Pvu*II-*Pst*I	1 531	8	6	14	1	3	4
*Not*I-*Eco*RV-*Hin*fI	1 600	14	5	19	4	2	6
*Not*I-*Bam*HI-*Hin*fI	1 786	12	13	25	3	2	5
*Not*I-*Pst*I-*Pvu*II	1 057	7	5	12	2	1	3
*Bss*HII-*Pvu*II-*Pst*I	2 545	6	12	18	2	5	7
*Bss*HII-*Bam*HI-*Eco*RV	777	14	8	22	2	4	6
*Bss*HII-*Bam*HI-*Eco*RI	1 269	5	5	10	0	2	2
Total	10 565	66	54	120	14	19	33

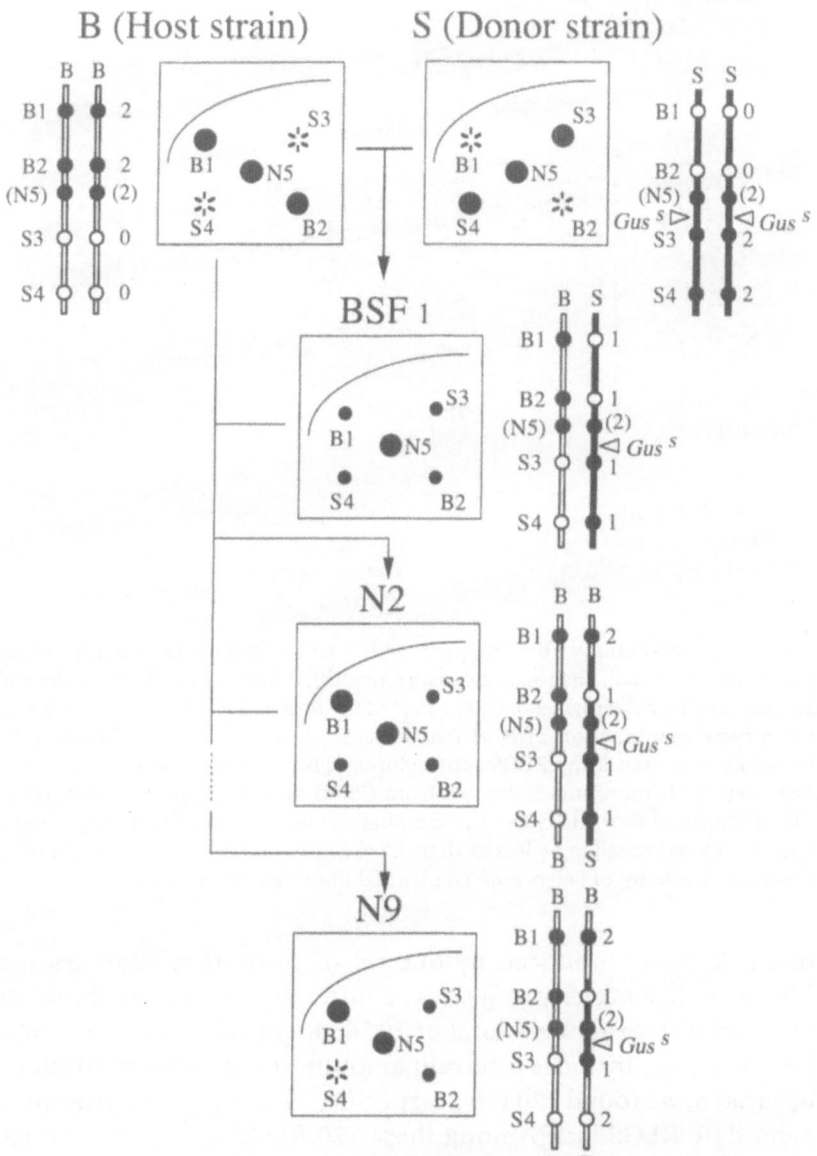

Fig. 5.3. Behaviro of RLGS spots mapped within *Gus^s* congenic region in the course of creation of a congenic strain. The map positions of five fictitious loci, *B1*, *B2*, *S3*, *S4*, and *N5*, are shown in the progenitor generation. *B1* and *B2* are B-specific spots. *S3* and *S4* are S-specific spots. *N5* is a nonpolymorphic spot. The *squares* represent RLGS profile. *Large dots* in the profile represent double intensity, which are homozygous for the presence of allele in the spots. *Small dots* represent single intensity, which are heterozygous for the loci. *The asterisk* * represents null intensity, which is homozygous for the absence of allele. *Vertical bold and white bars* depict chromosome derived from B (host strain) and S (donor strain), respectively. *Black and white circles* on the chromosomes represent presence and absence of allele in the spots, respectively

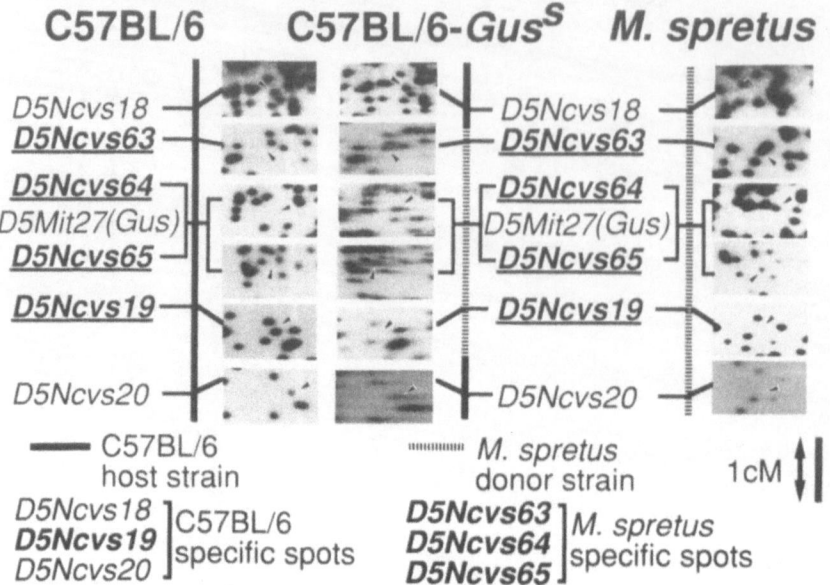

Fig. 5.4. Representative spots mapped within *Gus*ˢ congenic region after screening 10 565 RLGS loci. Each spot locus was magnified from an original RLGS profile and marked by an arrowhead. Only a part of the targeted RLGS loci detected using an enzyme combination of *Not*I-*Pvu*II-*Pst*I is shown, together with the established genetic map using RLGS spot mapping [14]. The *vertical bold and dotted lines* depict chromosome 5 derived from C57BL/6 and *M. spretus*, respectively. The underlined four RLGS loci are members of targeted spot loci. The congenic region was estimated to be less at than 4.2 cM interval [16]. (By copyright of The National Academy of Science of The United States of America)

of RLGS spots produced by one set of restriction enzymes were identified in each gel profile. Seven different enzyme combinations were used to examine a total of 10 565 loci (Table 5.1). This number, however, includes a certain amount of redundancy. With this approach, we found 120 (66 B6-specific, 54 Sp-specific) provisional candidate RLGS loci. Among these 120 RLGS loci, 33 (14 B6-specific, 19 Sp-specific) were mapped to the *Gus* region employing the BSS backcrosses [16]. Figure 5.4 demonstrates an example of a B6-specific spot (*D5Ncvs19*) that was missing from the congenic strain and *M. spretus*-specific spots (*D5Ncvs63*, *D5Ncvs64*, *D5Ncvs65*) that were present. In combination with the information of the genetic map of these RLGS loci (*D5Ncvs18*, *D5Ncvs20*), the *M. spretus* contribution to the congenic B6-*Gus*ˢ region was estimated to span an interval less than 4.2 cM [16].

Thirty three RLGS loci out of 120 provisional candidate RLGS loci were found that correctly mapped within a congenic region on chromosome 5.

The RLGS profiles of these 7 sets of enzyme combinations required a total of 21 RLGS gels. This task was completed with 4 days of laboratory work using the newly developed multiplex electrophoretic system RLGS Ver. 1.8 [17]. Hence, the loci identified in this study provide an average density of 1 locus/0.13 cM. This is well within the average physical size of nearly all large fragment clones. In the light of what would be expected from the total number of *Not*I sites (2380 sites) and *Bss*HII sites (about 9000 sites, our unpubl. data) estimated to be in the mouse genome, the number of loci identified for this congenic region is plausible. The basis for the identification of RLGS loci is endlabeled restriction fragments that typically result from overlapping cleavages. Hence, both sides of the cleavage site are endlabeled and, additionally, we assume both parental chromosomes to carry the same cleavage site. Consequently, there exist four possible cleavage products associated with each landmark site in the genome. However, if the rate of RLGS variation is about 50% between B6 and *M. spretus*, the probability that both sides of a landmark cleavage will be variant is 0.25. Moreover, only a portion of the total number of endlabeled restriction fragments for any particular enzyme set is identified with each RLGS profile. It is, therefore, possible to derive an estimate for identifying allelic RLGS fragments. Taken together, it is probable that some of the identified RLGS spots are either allelic or represent the same restriction landmark. However, the total number of independent loci still represents a density of at least 0.18 cM.

Those spots that did not map within the target region were false positive. A major reason for this apparently large number is that the congenic strain employed still contained a significant level of *M. spretus* genome that was independent of the *Gus* locus on chromosome 5. As was described in Okazaki et al. (1995) [16], we can estimate that the amount of remaining (R) donor genome is:

$$Rn = (0.5)^{n-1} \times (LE_{UL} - LE_L) \text{ cM},$$

where n is the number of generations backcrossed, LE is the length of the total genetic map (1600 cM) and the subscript letters of UL in LE_{UL} and L in LE_L are unlinked and linked donor genetic materials, respectively. Given nine backcross generations, we can estimate 6.25 cM of unlinked genome remaining, or about double the interval mapped in the chromosome five congenic region. Hence, the expected number of unlinked loci, 66 (double of 33), does not greatly differ from the 87 (120–33) that we observed.

RLGS is an efficient method for identifying the donor strain genes that are present in a congenic strain and it provides a direct method of determining the occurrence of both linked and unlinked donor genotype that remains in the final congenic strain. This method of assessment will become increasingly valuable in characterizing congenic strains constructed in accelerated breeding schemes that utilize expanded breeding populations in early generations to identify parents with reduced levels of donor genotype across the genome [18]. The RLGS method will be particularly helpful in analyzing selected congenic strains to check for unlinked donor genotypes missed by other marker methods.

5.3
Recombinant Inbred Strain Analysis

5.3.1
BXD RI Strains

Recombinant inbred (RI) strains are an ideal system for the analysis of RLGS loci. First, it is possible to analyze loci from both of the parental strains at the same time as SDPs on the basis of the locus's presence or absence. Hence, even though the number of variant loci is higher between laboratory strains and divergent species than between inbred strains, the total number of loci mapped is nearly equal to the analysis of the dominant markers segregating in an intersubspecific backcross. Second, an established database of reference loci, previously mapped in these strains, provides a direct assignment of these loci to chromosomal regions in the genetic map.

The variation between the inbred strains D2 and B6 is less than that observed between B6 and more divergent species. It offers, however, the important opportunity of utilizing RI strains to analyze the segregation of both parental strains in the same gels. Moreover, the cumulative genetic database for the 26 BXD RI strains provides a set of strain distribution patterns that identify

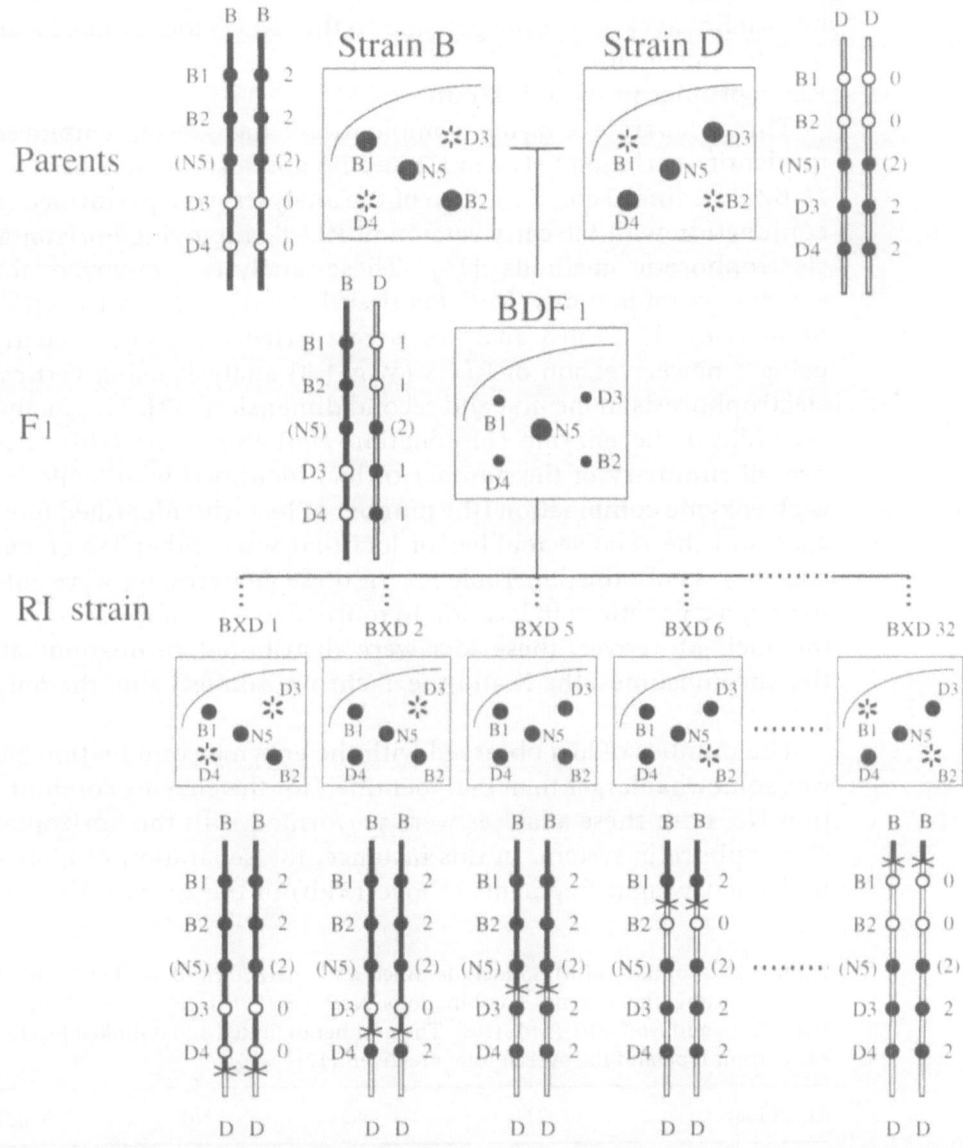

Fig. 5.5. Behavior of five fictitious RLGS spots on the chromosome in the course of creating an RI strain. The map positions of the five loci, *B1*, *B2*, *D3*, *D4*, and *N5*, are shown in the progenitor generation. *B1*, *B2*, and *D3*, *D4* are B-specific and D-specific spots, respectively. *N5* is a nonpolymorphic spot. The *squares* represent the RLGS profile. *Large dots* in the profile represent double intensity, and are homozygous for the presence of allele in the spots. *Small dots* in the F_1 generation represent single intensity, which are heterozygous for the loci. The *marks* * represent null intensity, and are homozygous for the absence of allele. The *vertical bold and white bars* depict chromosome derived from strain B and D, respectively. *Black and white circles* on the chromosomes represent presence and absence of allele respectively, in the spots. *Cross* represents a recombination event

known linkages across the genome so that RLGS loci can be localized to specific linkages. Figure 5.5 shows the schematic figure of RLGS profiles in BXD RI strain.

Three separate enzyme combinations have been employed to identify variation between D2 and B6 and also to yield SDP for 26 BXD strains (Table 5.2). Two of the analyses were performed in conjunction with the early version of RLGS employing horizontal electrophoretic methods [14]. These analyses employed the enzyme combinations NotI-PvuII-PstI (Na) and NotI-EcoRV-MboI (Nc). The third analyses were carried out more recently using a newer version of RLGS (Ver. 1.8) analysis using vertical electrophoresis in the first and second dimension [17]. This analysis utilized the enzyme combination NotI-PstI-PvuII (Nd). The overall summary of the number of loci identified with SDPs for each enzyme combination (the number of loci with identified linkages and the relative number of loci that were either D2 or B6-specific) is contained in Table 5.2. In these analyses, we were able to assign a genetic map location to more than 93% of the segregating loci. Moreover, these loci were distributed throughout all the chromosomes, the X and the Y chromosomes being the only exceptions.

The number of loci observed with the enzyme combination Na was somewhat larger than that identified for the enzyme combination Nc. Both these analyses were performed with the horizontal electrophoretic system. In this instance, the separation of higher molecular weight fragments (above 15 kb) in the agarose first di-

Table 5.2. Summary of RLGS loci identified and characterized as SDPs in BXD analyses using the enzyme combinations NotI-PvuII-PstI (Na), NotI-EcoRV-MboI (Nc), and NotI-PstI-PvuII (Nd). The number of linked and unlinked loci for each strain type and the overall total are given [17]

RLGS loci	Na	Nc	Nd	Total
B6 loci linked	135	97	98	330
unlinked	7	7	8	22
D2 loci linked	139	88	97	324
unlinked	15	4	6	25
Total loci linked	274	185	195	654
unlinked	22	11	14	47
Total loci	296	196	209	701

mension was not particularly effective, and the lower number of variant spots reflects the larger average size of *NotI-EcoRV* fragments compared with *NotI-PvuII* fragments. RLGS Ver 1.8 was used to carry out the analysis on the Nd restriction combination. It is not immediately clear, however, why the number of segregating loci was lower for this combination than in the case of the loci identified in Na. The enzyme combination Nd is representative of a reversal of the order of the second and third enzymes compared with Na. It is plausible that the relative frequency of *PvuII* cleavage is lower than that of *PstI* in the mouse genome. Therefore there is a lower probability of reducing *NotI-PstI* fragments into end-labeled fragments that will separate in the acrylamide second dimension of electrophoresis.

5.3.2
Genetic Analysis in BXD RI Strains

We analyzed the DNA from BXD RI strains with three enzyme combinations Na, Nc, and Nd (Table 5.2). Localization of spot loci on the RLGS profile of Nd is illustrated in Fig. 5.6. We identified both B6- and D2-specific loci in each analysis with 142, 104, and 106 B6-specific loci in enzyme combinations Na, Nc, and Nd, respectively, and 154, 92, and 103 D2-specific loci also in the same combinations. The SDPs for the loci of Nd (*NotI-PstI-PvuII*) are shown in Table 5.3 as an example.

Quite possibly, some of the D2- and B6-specific loci with coincident SDPs are, in fact, alleles of the same locus. In some cases these loci may be associated with RLGS spots that have comparable physical properties indicated in the second dimension of electrophoresis by a similarity of area code locations (Fig. 5.7). For example, the loci *D3Rik66* and *D7Rik71* identify spots NdB24 and NdD245, respectively. Both of these spots localize to area code I3. Using this criterion, 172 pairs of loci or a total of 344 loci were identified that showed identical properties throughout the genome. Thus, 25% of the loci we identified in the BXD analyses could be allelic, thereby decreasing the total number of independent sites mapped to between 75 and 90% of the 473 loci distributed in the BXD maps.

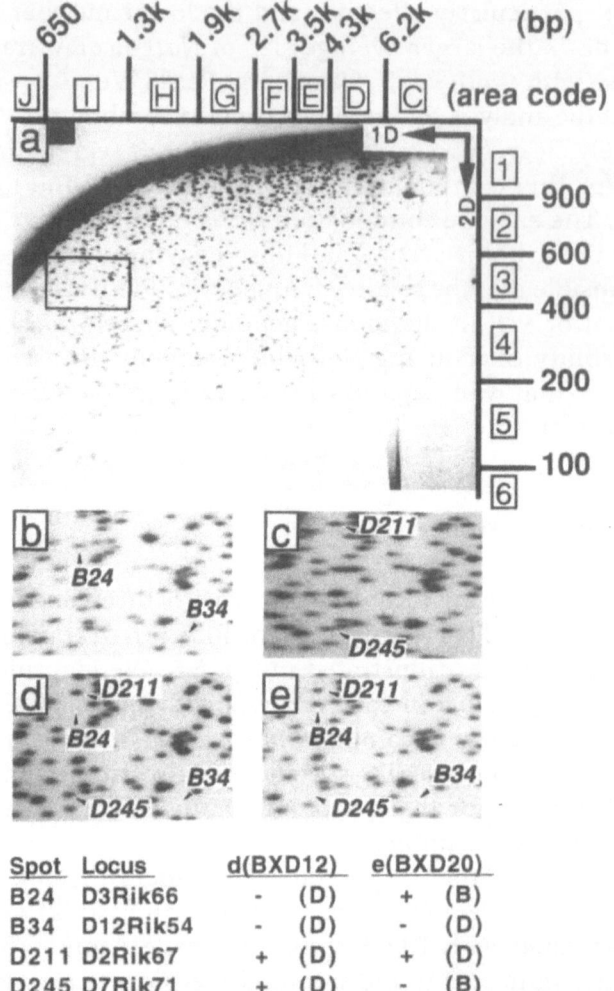

Spot	Locus	d(BXD12)	e(BXD20)
B24	D3Rik66	- (D)	+ (B)
B34	D12Rik54	- (D)	- (D)
D211	D2Rik67	+ (D)	+ (D)
D245	D7Rik71	+ (D)	- (B)

Fig. 5.6a–e. Localization of spot loci on the RLGS profile. **a** BDF_1 profile using enzyme combination *NotI-PstI-PvuII* is shown. Molecular size marker is shown on the periphery and area code is typed as C-J for the 1st-D and 1–6 for the 2nd-D according to the molecular size marker. B6- and D2-specific spots were identified by comparing B6 and D2 patterns with BDF_1 pattern and numbered (not shown). Part of the RLGS pattern shown in *closed square* in **a** is magnified and demonstrated as: **b** B6; **c** D2; **d** BXD12; **e** BXD20. *B24*, *B34*, and *D211*, *D245* show the representative spots of B6- and D2-specific spots, respectively. Segregation of these spots in the two strains BXD12 and BXD20 is shown in **d** and **e**, and the presence or absence of these spots is given as + or −. The presence of a B(D)-specific spot implies that the locus is derived from B6(D2), and the absence of a B(D)-specific spot means derived from D2(B6). [17] (By permission of Academic Press)

Table 5.3. List of RLGS loci typed in the BXD analyses showing the locus identification number and area code in the RLGS profile produced by using the enzyme combination of *NotI-PstI-PvuII*. The SDP for each of these loci in the BXD series is given. Area code is determined by the molecular size ranges in the subdivisions of the first and second dimension of the electrophoretic separation of RLGS [17]

Locus	Spot	Area Code	1	2	5	6	8	9	1 1	1 2	1 3	1 4	1 5	1 6	1 8	1 9	2 0	2 1	2 2	2 3	2 4	2 5	2 7	2 8	2 9	3 0	3 1	3 2
D1Ncvs102	D 270	D1	B	B	D	D	D	B	D	D	B	B	D	D	D	D	D	D	B	B	D	D	B	D	B	B	B	B
D1Ncvs90	B 51	H4	B	B	D	D	B	B	B	B	D	B	D	B	B	B	D	B	B	D	D	B	B	D	D	B	B	D
D1Ncvs91	D 202	G1	B	B	D	D	B	B	D	B	D	B	D	B	B	B	D	B	D	D	D	B	D	D	D	B	B	D
D1Ncvs92	B 6	G1	B	B	D	D	B	B	D	B	D	B	D	B	B	B	D	B	D	D	D	B	D	D	D	B	B	D
D1Ncvs94	D 302	F4	B	B	D	D	B	B	B	B	D	B	D	B	B	B	D	B	B	D	D	B	B	D	D	B	B	D
D1Ncvsd95	D 239	H4	B	B	D	D	B	B	D	B	D	B	D	B	B	B	D	B	B	D	D	B	B	D	D	B	B	D
D1Ncvs96	B 25	I3	B	B	D	D	B	B	D	B	D	B	D	B	B	B	D	B	B	D	D	B	D	U	D	B	B	D
D1Ncvs103	B 54	G4	B	B	D	D	B	B	D	B	D	B	D	D	B	B	D	B	B	D	D	B	D	D	D	B	B	D
D1Ncvs88	B 4	G1	B	B	D	U	B	D	B	B	D	B	B	D	B	B	D	B	D	D	D	D	D	D	D	B	B	D
D1Ncvs89	D 206	H2	B	B	D	B	B	D	B	B	D	B	D	D	B	B	D	B	D	D	D	D	D	D	D	B	B	D
D1Ncvs104	D 308	F5	B	B	D	B	B	D	B	B	D	B	D	B	B	B	D	B	D	D	D	D	D	D	D	B	B	D
D1Ncvs97	D 288	G1	B	B	D	B	B	D	B	B	D	B	D	D	B	B	D	B	D	D	D	D	D	D	D	B	B	D
D1Ncvs98	D 297	C3	B	B	D	B	B	D	D	B	D	B	B	D	B	B	D	B	D	D	D	D	D	D	D	B	B	U
D1Ncvs99	D 296	D3	B	B	D	B	B	D	D	B	D	B	B	D	B	B	D	B	D	D	D	D	D	D	D	B	B	D
D1Ncvs86	B 97	C3	B	B	D	B	B	U	D	B	D	B	B	D	B	B	D	B	D	D	D	D	D	D	D	B	B	U
D1Ncvs87	B 89	D3	B	B	D	B	B	D	B	B	D	B	B	D	B	B	D	B	D	D	D	D	D	D	D	B	B	D

Table 5.3. *Continued*

Locus	Spot	Area Code	32	31	30	29	28	27	25	24	23	22	21	20	19	18	16	15	14	13	12	11	9	8	6	5	2	1
D1Ncvs100	B 40	D3	B	B	D	D	D	D	D	D	D	D	B	B	B	B	B	B	B	D	B	D	D	B	B	D	B	B
D1Ncvs101	B 55	I5	B	B	D	D	D	D	D	U	D	D	D	B	B	B	B	D	B	D	B	B	D	D	U	D	B	B
D2Ncvs71	B 36	H3	D	B	B	D	U	B	B	D	U	D	D	D	B	B	B	B	D	D	B	B	U	D	B	D	B	D
DD2Ncvs72	D 227	G3		D	D	B	B	B	D	D	B	D	D	B	B	B	D	B	D	B	D	B	D	D	B	B	B	D
D2Ncvs73	D 203	G1	D	D	D	B	B	B	D	D	B	D	D	B	B	B	D	B	D	B	D	B	D	D	B	D	D	D
D2Ncvs74	B 38	G3	D	D	D	B	B	B	D	D	D	D	D	B	D	B	D	D	D	B	B	D	B	D	B	B	D	D
D2Ncvs70	B 8	G2	D	D	D	B	B	B	D	D	D	D	D	B	D	B	D	D	D	B	B	D	B	D	B	B	D	D
D2Ncvs75	D 220	G2	D	B	D	D	D	D	B	D	D	B	D	D	D	D	D	D	B	B	B	D	D	D	B	B	B	B
D2Ncvs63	B 21	I2	D	B	D	D	D	D	B	D	D	B	D	D	D	D	D	D	B	B	B	D	D	D	D	B	B	B
D2Ncvs64	D 207	I2	B	B	D	D	D	D	B	D	D	B	D	D	D	D	D	D	B	B	D	D	D	D	D	B	B	B
D2Ncvs65	D 316	E4	B	B	D	D	D	D	B	D	D	B	D	D	D	D	D	B	B	B	D	D	D	D	D	B	B	B
D2Ncvs66	B 84	E2	D	D	D	D	B	D	D	D	D	B	D	D	D	D	B	B	B	D	D	D	D	B	U	B	D	D
D2Ncvs67	D 211	I3	D	D	D	D	B	D	D	D	D	B	D	D	D	D	U	B	B	D	D	D	D	U	B	B	D	D
D2Ncvs68	B 64	G1	D	D	D	D	U	D	D	D	D	B	D	D	D	D	B	U	B	D	D	D	D	D	B	B	D	D
D2Ncvs69	D 205	H2	D	D	D	D	B	D	D	D	B	B	D	D	D	D	U	D	B	D	D	D	D	D	B	B	D	D
D3Ncvs64	B 72	F1	D	D	D	D	D	D	B	D	B	B	D	D	D	D	B	U	B	D	D	D	D	D	B	B	D	D
D3Ncvs65	B 77	F1	B	D	B	B	D	D	B	D	B	D	D	B	D	D	B	D	D	B	D	B	D	D	U	B	D	B
D3Ncvs66	B 24	I3	B	D	D	D	U	D	B	B	B	D	D	B	B	D	D	B	D	B	D	D	D	B	B	B	D	D
D3Ncvs53	D 210	I2	B	B	D	B	D	D	B	B	B	D	D	B	B	D	B	D	B	B	D	D	D	D	B	B	D	B
D3Ncvs54	D 264	D2	B	B	D	B	D	B	D	B	D	D	D	B	B	B	B	D	B	B	D	D	D	D	B	D	B	B

Clone	D3Ncvs55	D3Ncvs56	D3Ncvs57	D3Ncvs58	D3Ncvs59	D3Ncvs60	D3Ncvs61	D3Ncvs62	D3Ncvs63	D4Ncvs124	D4Ncvs125	D4Ncvs127	D4Ncvs123	D4Ncvs102	D4Ncvs103	D4Ncvs104	D4Ncvs105	D4Ncvs128	D4Ncvs106	D4Ncvs107	D4Ncvs108	D4Ncvs109	D4Ncvs110
Spot	D 307	D 213	D 310	D 262	D 258	B 117	B 42	D 284	D 301	D 291	D265	D 238	B 59	D 208	B 20	B 111	B 90	D 276	D 219	B 113	D 314	B 68	B 19
Code	D5	I3	F4	E2	F3	F3	F4	F2	F4	G2	D2	G5	I5	H2	H2	D4	D3	F1	H2	G4	F2	F2	H2
	U	B	D	D	D	D	D	D	D	D	D	D	B	B	U	U	B	B	B	B	B	B	B
	B	B	B	B	B	B	B	B	B	B	B	B	B	B	B	B	B	B	B	B	B	B	B
	D	D	B	B	B	B	B	B	B	D	D	D	D	D	D	D	D	D	D	D	D	D	D
	B	B	B	B	B	B	B	B	B	B	B	B	B	B	B	B	B	B	B	B	B	B	B
	D	D	D	D	D	D	D	D	D	B	B	B	B	B	B	B	B	B	B	B	D	D	D
	B	B	B	B	B	B	B	B	B	B	D	D	B	B	B	B	D	D	D	D	D	D	D
	D	D	D	D	D	D	D	D	D	D	D	D	D	D	D	D	D	D	D	D	D	D	D
	B	B	B	B	B	B	B	B	B	D	D	D	D	D	D	D	D	D	D	D	D	D	D
	B	B	B	B	D	D	D	D	D	B	B	B	B	B	B	B	B	B	B	B	B	B	B
	D	D	D	D	B	B	B	B	B	B	B	B	D	D	D	D	D	D	B	B	B	D	D
	D	D	D	D	D	D	D	D	D	D	B	B	D	D	D	D	D	B	D	D	D	D	D
	B	B	D	D	D	D	D	D	D	B	B	B	B	B	B	B	B	B	B	B	B	B	B
	B	B	B	B	D	D	D	D	D	D	D	D	D	B	B	B	B	B	B	B	D	D	D
	B	B	B	B	D	D	D	D	D	D	D	B	B	B	B	B	B	B	B	B	D	D	B
	B	B	B	B	B	B	B	B	B	B	D	D	B	B	B	B	B	B	B	B	B	B	B
	D	D	D	D	B	B	B	B	B	D	D	B	B	B	B	B	D	D	D	D	D	D	D
	B	B	B	B	B	B	B	B	B	B	B	B	B	B	B	B	B	B	B	B	B	B	B
	B	B	D	D	D	D	D	D	D	D	D	D	D	D	D	D	D	D	D	D	D	D	D
	B	B	D	D	D	D	D	D	D	B	B	B	B	B	B	B	B	B	B	B	B	B	B
	D	D	D	D	D	D	D	D	D	D	D	D	B	B	B	B	B	B	B	B	B	B	B
	B	B	B	B	B	B	B	B	B	D	B	B	B	B	B	U	B	B	D	U	D	D	D
	D	D	D	D	D	D	D	D	D	B	B	B	B	B	B	B	B	B	B	B	B	B	B
	B	B	B	B	B	B	B	B	B	B	B	D	D	B	B	U	B	B	B	B	B	B	B
	B	B	B	B	B	B	B	B	B	B	B	B	B	B	B	B	B	B	B	B	B	B	B
	B	B	D	D	B	D	D	D	D	B	B	D	D	D	B	B	B	D	B	B	D	B	B
	B	B	B	B	B	B	B	B	B	B	B	B	B	B	B	B	B	B	B	B	B	B	B
	B	B	B	B	B	B	B	B	B	B	D	D	D	D	D	D	D	D	D	D	D	D	D

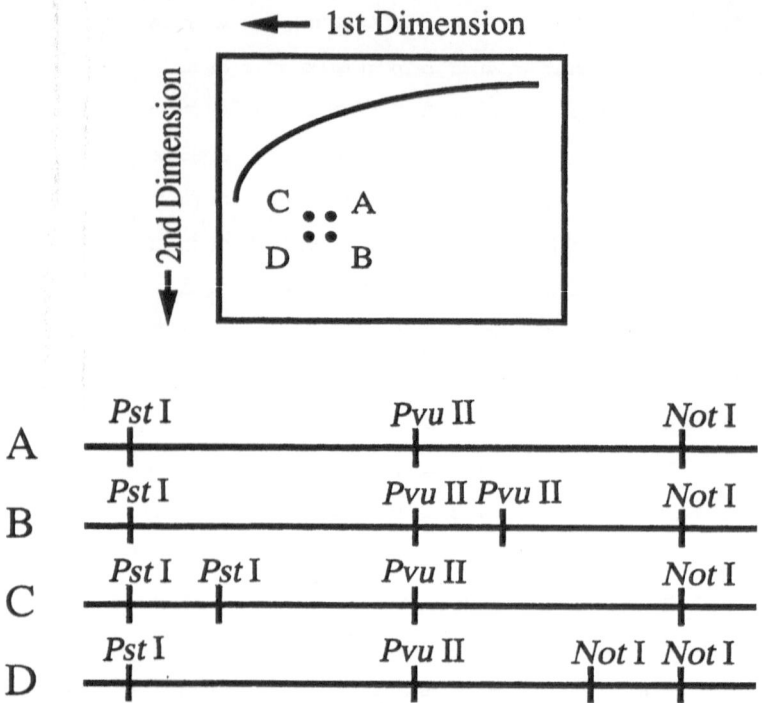

Fig. 5.7. Simple explanation of behavior of allelic spots (paired loci in the RLGS map) with site polymorphism. The *squares* represent RLGS profile of restriction enzyme combination, *Not*I, *Pst*I, and *Pvu*II for E_L, E_B, and E_C, respectively, as an example. Spots *A*, *B*, *C*, and *D* are derived from the DNA fragment as shown with the restriction map A, B, C, and D, respectively

5.4
Backcross Analyses

To identify and map RLGS loci, interspecific backcrosses have been employed. In the cross (B6 × *M. spretus*)F$_1$ × *M. spretus* (BSS), 72 progeny were analyzed for the segregation of B6 and *M. spretus* loci in the second cross. In this cross, we characterized the segregation of centromeric heterochromatin differences between B6 and *M. spretus* for the progeny using FISH analysis of the major satellite sequences. We identified the segregation of the B6 centromeric satellite as a single locus for each chromosome using fluorescence banding to karyotype mitotic chromosomes [19]. As primary markers for the construction of the genetic maps, the centromeric heterochromatin loci of this cross were used. After determining the progeny distribution patterns (PDPs) for each B6 locus for the cross, they were entered into the file of a computer software, named Map Manager [20] (Contact Dr. Kenneth F. Manly: Department of Cellular and Molecular Biology, Roswell Park Cancer Institute, Buffalo, NY 14263-0001, USA. Map Manager home page: http://mcbio.med.buffalo.edu/mapmgr.html. Map Manager email: mapmgr@mcbio.med.buffalo.edu) with the centromeric heterochromatin segregation pattern. We built chromosome maps by identifying the closest linkage to each centromeric region and then adding additional loci to each chromosome with increasingly larger recombination frequencies. Gene order was ascertained on the principle of minimizing multiple recombination events within each of the indicated haplotypes. In the cases (particularly those involving a single locus) where a locus position with the minimal recombination produced multiple recombination events in the haplotype, all of the PDPs were reexamined for that locus to determine whether an error in genotypic determinations had occurred. In those cases where reexamination of the autoradiographic signals did not result in a change in the original genotypic score, that locus was removed from the ordered genetic map. The schematic figure of RLGS spot mapping method utilizing BSS backcross progeny is illustrated in Fig. 5.8.

5.4.1
Analysis of Dominant Loci

In developing a genetic map, our methodology was largely successful for the BSS cross. However, genetic maps for two chromosomes, Chr 1 and Chr 15, were initially shorter than expected. Additional anchor loci were then added to the data set to cover these regions. The final genetic map of RLGS loci in the BSS cross included 276 ordered B6-specific loci dispersed among all the chromosomes, covering more than 1341 cM of cumulative genetic map distance. We mapped additional 47 loci to different chromosomes with an LOD score of 3 or greater. In these instances, their positions can be estimated. They are nevertheless not included, in haplotype analysis of recombinational distances on the chromosomes. A summary of the number of loci mapped and the recombinational distances can be found in Table 5.4; the overall genetic maps have previously been published [14].

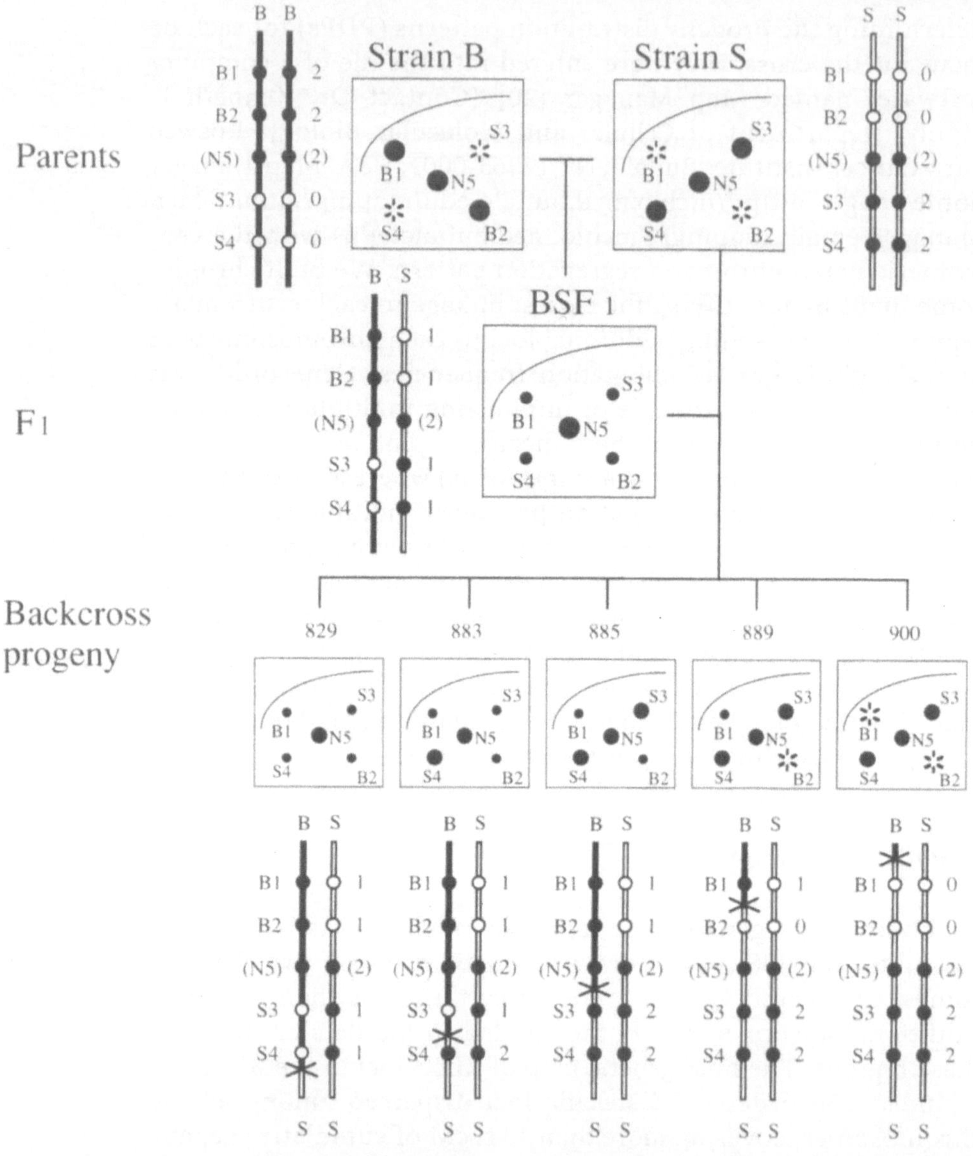

Fig. 5.8. Behavior of five fictitious RLGS spots on the chromosome in the course of creation of backcross progeny. The map positions of the five loci, *B1*, *B2*, *S3*, *S4*, and *N5* are shown in the progenitor generation. *B1*, *B2*, and *S3*, *S4* are B-specific, and S-specific spots, respectively. *N5* is nonpolymorphic spot. The *squares* represent the RLGS profile. *Large dots* in the profile represent double intensity, and are homozygous for the presence of allele in the spots. *Small dots* represent single intensity, and are heterozygous for the loci. The *marks* * represent null intensity, and are homozygous for the absence allele. The *vertical bold and white bars* depict chromosome derived from strain B and S, respectively. *Black and white circles* on the chromosomes represent presence and absence of allele in the spots, respectively. *Cross* represents a recombination event

Table 5.4. Summary of loci identified and mapped in (B6 × *M. spretus*) F₁ × *M. spretus* (BSS) including B-specific dominant loci and S-specific codominant loci[a] [35]

Cross	Loci ordered	Loci mapped	Loci unlinked	Total loci	Total cM	Recombination
BSS – B loci	276	47	2	325	1341	985
BSS – S loci	165	64	21	250		

[a] The total recombinational distance covered in each map and the number of recombination events recorded are also shown.

5.4.2
Analysis of Codominant Loci

The initial backcross analyses were based upon the segregation of the presence or absence of an RLGS spot. In theory, the quantitative levels of endlabeling should also segregate in the backcross progeny either as a single allelic copy with half-intensity or as a double copy of the locus with full-intensity labeling (in comparison to nonsegregating loci with full intensity). We based these analyses on identifying *M. spretus* (S)-specific alleles in the RLGS profile of (B6 × *M. spretus*) F₁ × *M. spretus* (BSS) progeny and by tracking the intensity of these loci in the BSS backcross progeny [15]. With independent autoradiographic film readers, the S-specific allele distribution patterns were established. These were subsequently compared with the segregation of the B-specific alleles.

A total of 165 S-loci that could be ordered within haplotypes of all the chromosomes were analyzed in this manner. We identified additional 64 loci that could be linked to specific chromosome regions with LOD scores of 3 or greater. However, there were too many double recombinant haplotypes to determine the logically best-fitting gene order (Table 5.4). The addition of the S-specific loci to the BSS genetic map did not change the cumulative amount of recombination across the genome of 1341 cM.

The number of segregating loci that could be analyzed for each backcross progeny in a single gel nearly doubled with the addition of the codominant S-specific loci to the BSS analysis. In some cases, it was possible to integrate additional B-specific loci into the overall genetic map when we added the S-specific loci. Practically speaking, the additional loci create an average map density of one marker for each 2.5 cM across the total genetic map. These results can be used to significant advantage in the analysis of the loss of heterozygosity (LOH) in experimental tumors involving crosses of laboratory strains bearing transgenes of tissue-specific, oncogene-driving promoters ([21]–[23] see Chap. 7). In theory, genetic markers present at this density will greatly aid both positional cloning work [24] and the analysis of complex or quantitative trait loci.

Table 5.5 illustrates an example of the integration of nine S-loci into the previously characterized B-loci and other anchor markers of chromosome 1 for 72 BSS progeny. In these data, 17 B-specific loci are identified as *D1Ncvs* numbers. Four chromosome 1 anchor loci are shown, including the centromeric heterochromatin locus (*Hcl*), two *D1Mit* loci, and the functional gene, *Acrg*. The addition of *D1Mit17* extends the distal genetic map an additional 30 cM beyond *D1Ncvs15* and accounts for closely linked loci that are coincident with *D1Mit17* in their segregation patterns. A total of 26 RLGS *Not*I loci have been mapped to the chromosome. Seven of the loci are dispersed over the proximal 50 cM of the chromosome. However, 17 are clustered within a 20-cM region in the middle of the chromosome and additional 3 loci are coincident with the distal anchor marker, *D1Mit17*. It is difficult to determine whether this distribution reflects

(1) a concentration of genes with CpG islands in the middle regions of the chromosomes, or

(2) a consequence of the limited sample size of BSS progeny (Table 5.6)

The segregation of these loci, the recombination fraction between loci, and the estimated map distance from the file of a computer software, Map Manager [20], are also shown along with the cumulative genetic map for the chromosome (Fig. 5.9).

5.4.3
Genetic Analysis of RLGS Loci in the BSS Backcross

The PDPs of 575 RLGS loci were analyzed for 72 BSS progeny. Of these, the PDPs of 441 loci demontrated linkages and defined

Chr 1

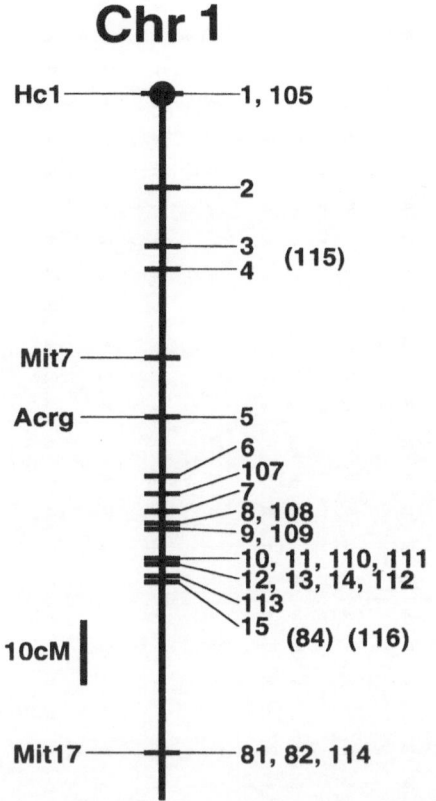

Fig. 5.9. Cumulative genetic map for mouse chromosome 1 with the centromeric heterochromatin marker, *Hc1* and additional anchor loci. Dominant B6-specific spots are identified as *D1Ncvs1* through, *84* while codominant *M. spretus*-specific spots are identified as *D1Rik105* through *116*. The locus *numbers* shown correspond to the *D1Ncvs* and *D1Rik* numbers. Loci shown *in parenthesis* are localized by the LOD score, but are not included in the ordered chromosome because of single locus double recombinations that make it impossible to order these loci unambiguously [35]. (By permission of VCH)

haplotypes for each of the chromosomes. The distribution of the ordered RLGS loci in the BSS series is shown in Table 5.7. These loci have been placed on chromosomal genetic maps primarily according to their association to the segregating, major satellite that determined the centromeric heterochromatin locus (*Hc*) for each chromosome using karyotype analysis and FISH. We have added additional 24 reference loci to the maps of specific chromosomes to aligh and extend the genetic maps. Among these loci are SSR loci and some other functional genes that can be identified on

Table 5.5. Progeny distribution patterns of C57BL/6 (B) and *M. spretus* (S) alleles segregating for Chr 1 in the BSS backcross. The loci are listed in the best-fit order and the progeny are aligned in a haplotype sort from the Map Manager program beginning with B and S parental types that are non-recombinant and showing single and double recombinants in descending order. D1Ncvs- loci identify B-specific loci while D1Rik- loci identify s-specific loci [35]

Locus	Progeny
Hc1	
D1Ncvs1	
D1Rik105	
D1Ncvs2	
D1Ncvs3	
D1Ncvs4	
D1Mit7	
D1Ncvs5	
Acrg	
D1Ncvs6	
D1Rik107	
D1Ncvs7	
D1Rik108	

Locus																																
DINcvs8	B	s	s	B	s	B	B	s	B	B	B	B	s	B	s	B	B	s	B	s	B	B	s	B	s	B	B	s	B	s	B	s	B
D1Rik109	B	s	s	B	s	B	B	s	B	B	B	B	s	B	s	B	B	s	B	s	B	B	s	B	s	B	B	s	B	s	B	s	B
DINcvs9	B	s	s	B	s	B	B	s	B	B	B	B	s	B	s	B	B	s	B	s	B	B	s	B	s	B	B	s	B	s	B	s	B
DINcvs10	B	s	s	B	s	B	B	s	B	B	B	B	s	B	s	B	B	s	B	s	B	B	s	B	s	B	B	s	B	s	B	s	B
D1Rik110	B	s	s	B	s	B	B	s	B	B	B	B	s	B	s	B	B	s	B	s	B	B	s	B	s	B	B	s	B	s	B	s	B
DINcvs11	B	s	s	B	s	B	B	s	B	B	B	B	s	B	s	B	B	s	B	s	B	B	s	B	s	B	B	s	B	s	B	s	B
D1Rik111	B	s	s	B	s	B	B	s	B	B	B	B	s	B	s	B	B	s	B	s	B	B	s	B	s	B	B	s	B	s	B	s	B
DINcvs13	B	s	s	B	s	B	B	s	B	B	B	B	s	B	s	B	B	s	B	s	B	B	s	B	s	B	B	s	B	s	B	s	B
DINcvs14	B	s	s	B	s	B	B	s	B	B	B	B	s	B	s	B	B	s	B	s	B	B	s	B	s	B	B	s	B	s	B	s	B
D1Rik112	B	s	s	B	s	B	B	s	B	B	B	B	s	B	s	B	B	s	B	s	B	B	s	B	s	B	B	s	B	s	B	s	B
DINcvs12	B	s	s	B	s	B	B	s	B	B	B	B	s	B	s	B	B	s	B	s	B	B	s	B	s	B	B	s	B	s	B	s	B
D1Rik113	B	s	s	B	s	B	B	s	B	B	B	B	s	B	s	B	B	s	B	s	B	B	s	B	s	B	B	s	B	s	B	s	B
DINcvs15	B	s	s	B	s	B	B	s	B	B	B	B	s	B	s	B	B	s	B	s	B	B	s	B	s	B	B	s	B	s	B	s	B
D1Mit17	B	s	s	B	s	B	B	s	B	B	B	B	s	B	s	B	B	s	B	s	B	B	s	B	s	B	B	s	B	s	B	s	B
DINcvs82	B	s	s	B	s	B	B	s	B	B	B	B	s	B	s	B	B	s	B	s	B	B	s	B	s	B	B	s	B	s	B	s	B
DINcvs81	B	s	s	B	s	B	B	s	B	B	B	B	s	B	s	B	B	s	B	s	B	B	s	B	s	B	B	s	B	s	B	s	B
D1Rik114	B	s	s	B	s	B	B	s	B	B	B	B	s	B	s	B	B	s	B	s	B	B	s	B	s	B	B	s	B	s	B	s	B
No. of progeny	1	1	1	1	1	1	1	1	1	2	1	1	1	1	1	1	1	1	1	2	8	1	1	1	1	1	3	2	2	5	4	5	9

Table 5.6. Segregation and recombination between loci on BSS chromosome 1 [35]

Locus	Mat	Pat	X	N	X/N	S.E.	LOD
Hc1	44	28					
			0	65	0.00	0.00	19.6
D1Ncvs1	40	25					
			0	65	0.00	0.00	19.6
D1Rik105	43	28					
			11	69	15.94	4.41	7.6
D1Ncvs2	45	25					
			7	69	10.14	3.63	10.9
D1Ncvs3	47	24					
			3	66	4.54	2.56	14.6
D1Ncvs4	42	25					
			9	67	13.43	4.17	8.7
D1Mit7	46	26					
			6	67	8.95	3.49	11.4
D1Ncvs5	40	27					
			0	67	0.00	0.00	20.2
Acrg	44	28					
			7	72	9.72	3.49	11.7
D1Ncvs6	43	29					
			3	71	4.23	2.39	16.0
D1Rik107	41	30					
			2	57	3.51	2.44	13.4
D1Ncvs7	40	18					
			1	57	1.75	1.74	15.0
D1Rik108	42	29					
			1	71	1.41	1.40	19.1
D1Ncvs8	44	28					
			1	69	1.45	1.44	18.5
D1Rik109	40	29					
			0	69	0.00	0.00	20.8
D1Ncvs9	43	29					
			3	64	4.69	2.64	14.0
D1Ncvs10	35	29					
			0	63	0.00	0.00	19.0
D1Rik110	43	28					
			0	62	0.00	0.00	18.7
D1Ncvs11	39	24					
			0	60	0.00	0.00	18.1
D1Rik111	40	29					
			1	69	1.45	1.44	18.5
D1Ncvs13	43	29					
			0	67	0.00	0.00	20.2
D1Ncvs14	42	25					
			0	64	0.00	0.00	19.3

Table 5.6. *Continued*

Locus	Mat	Pat	X	N	X/N	S.E.	LOD
D1Rik112	40	29					
			0	61	0.00	0.00	18.4
D1Ncvs12	42	22					
			1	63	1.59	1.57	16.7
D1Rik113	41	30					
			1	71	1.41	1.40	19.1
D1Ncvs15	43	29					
			21	72	29.17	5.36	2.8
D1Mit17	34	38					
			0	70	0.00	0.00	21.1
D1Ncvs82	33	37					
			0	66	0.00	0.00	19.9
D1Ncvs81	32	36					
			0	66	0.00	0.00	19.9
D1Rik114	32	38					

the chromosomal maps (Fig. 5.10) Recombination among these loci resulted in an estimate of 1341 map units for the mouse genome. A total of 985 recombination events were identified in these analyses. Of these, 679 were single recombinant events (69%), 288 (144 × 2) were double crossovers (29%), and 18 (6 × 3) were triple recombinations (2%).

Additional 111 loci (linked but no haplotype – Table 5.7) have been assigned to chromosomes by linkage analysis and given a D-number for a specific chromosome. The location of these loci on the chromosome was determined using the highest LOD score (ranging from 4 to 10). Their order, however, could not be established unambiguously by haplotype analysis and they are therefore not included in the final estimation of chromosome size in the genetic map. These loci are given in parentheses to indicate the provisional nature of the localization on the genetic maps of chromosome (Fig. 5.10). Thus, there is a total of 596 loci in the BSS genetic map (552 RLGS loci + 20 *Hc* loci + 24 reference loci).

5.5
Coordinated Genetic Maps

The genetic maps derived from each of the different analyses were partially integrated by the use of common anchor loci between

Table 5.7. Chromosomal distribution of RLGS loci in the BSS cross[a] [35]

Chr	Anchor loci	RLGS loci						Total RLGS loci	Total mapped loci	RLGS paired loci	Haplotypes					Recombination events	Length (cM)
		RLGS loci, ordered			Linked loci, no haplotype						Nonrecombinant		Recombinant				
		B	S	Total	B	S	Total				Hetero-zygous	Homo-zygous	Single	Double	Triple		
1	4	17	10	27	1	2	3	30	34	6	9	5	38	18	2	80	112.81
2	3	19	15	34	5	8	13	47	50	6	7	15	34	15	1	67	90.84
3	2	9	5	14	5	5	10	24	26	4	10	13	42	7	0	56	73.33
4	1	20	11	31	9	7	16	47	48	8	15	11	32	14	0	60	77.56
5	3	21	11	32	3	5	8	40	43	6	19	14	29	9	1	50	65.69
6	3	11	7	18	1	1	2	20	23	2	16	13	29	14	0	57	75.08
7	1	23	10	33	2	3	5	38	39	8	15	9	41	6	1	56	73.37
8	1	18	7	25	2	4	6	31	32	4	18	10	34	10	0	54	71.39
9	1	17	11	28	2	2	4	32	33	6	12	13	40	7	0	54	73.05
10	2	10	15	25	3	2	5	30	32	6	11	12	40	9	0	58	89.48
11	4	22	18	40	1	2	3	43	47	6	16	10	33	13	0	59	71.69
12	1	12	3	15	2	6	8	23	24	0	16	24	27	5	0	37	51.51

Chr																	Genetic length
13	2	14	10	24	0	2	2	26	28	4	20	14	37	1	0	39	54.83
14	1	14	6	20	3	2	5	25	26	8	22	24	26	0	0	26	34.79
15	2	9	9	18	1	4	5	23	25	0	21	11	37	3	0	43	58.70
16	6	10	4	14	1	6	7	21	27	4	25	14	29	4	0	37	52.79
17	1	10	5	15	1	1	2	17	18	6	20	16	36	0	0	36	49.42
18	2	6	4	10	2	2	4	14	16	2	21	19	31	1	0	33	46.20
19	1	10	4	14	3	0	3	17	18	4	25	18	28	1	0	30	42.31
X	3	4	0	4	0	0	0	4	7	0	4	24	36	7	1	53	76.40
Subtotal	44	276	165	441	47	64	111	552	596	90	322	289	679	144	6	985	1341.24
Unlinked	0	0	0	0	2	21	23	23	23	0							
Total	44	276	165	441	49	85	134	575	619	90							

[a] This table indicates the number of RLGS loci mapped to each chromosome (Chr) with or without anchor loci, the number of recombination events caused on each chromosome, the number of RLGS loci with double recombination events at their both adjacent sites that could be located by linkage, and the number of paired loci estimated as a pair of alleles. The total genetic length of the chromosome is also represented in the rightmost column.

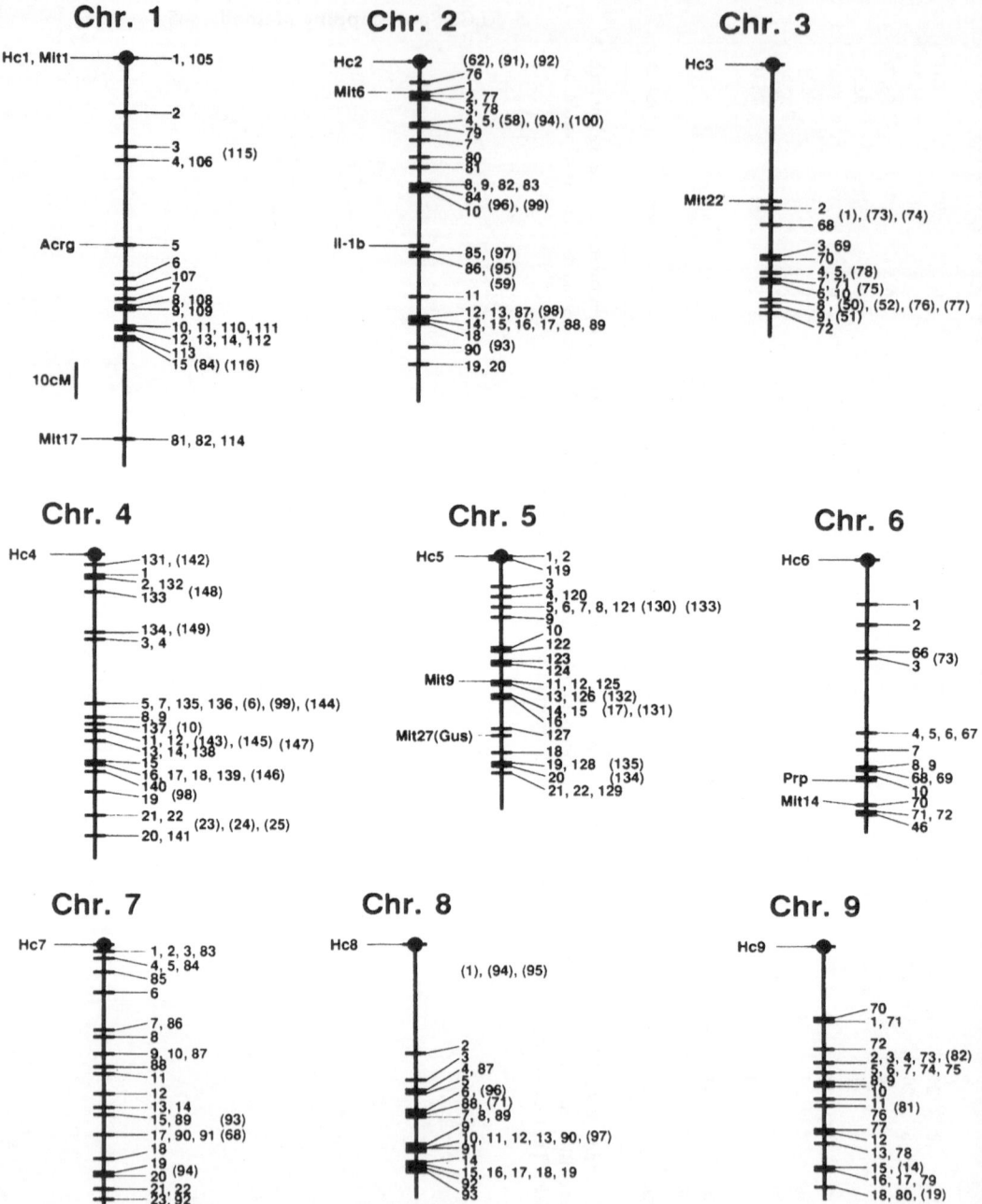

Fig. 5.10. Genetic map of 596 B- and S-specific *Not*I loci on the autosomes and X chromosomes of the mouse. The relative length of the BSS chromosomes was estimated from cumulative recombination events using the haplotype analysis function of the file of a computer software, named Map Manager. A single locus with double recombination events with neighboring loci was provisionally linked to a locus if it gave an LOD score of more than 3. However, these loci were eliminated from the haplotype analysis to estimate the genetic length of total genome. In this figure, these were represented by the number of *D-Ncvs-* or *D-Rik-* in parentheses. Anchor loci, such as *Mit* SSR loci, are shown on the *left* side. The RLGS loci are named *D-Ncvs* or *D-Rik* and are sequentially numbered, as described in the

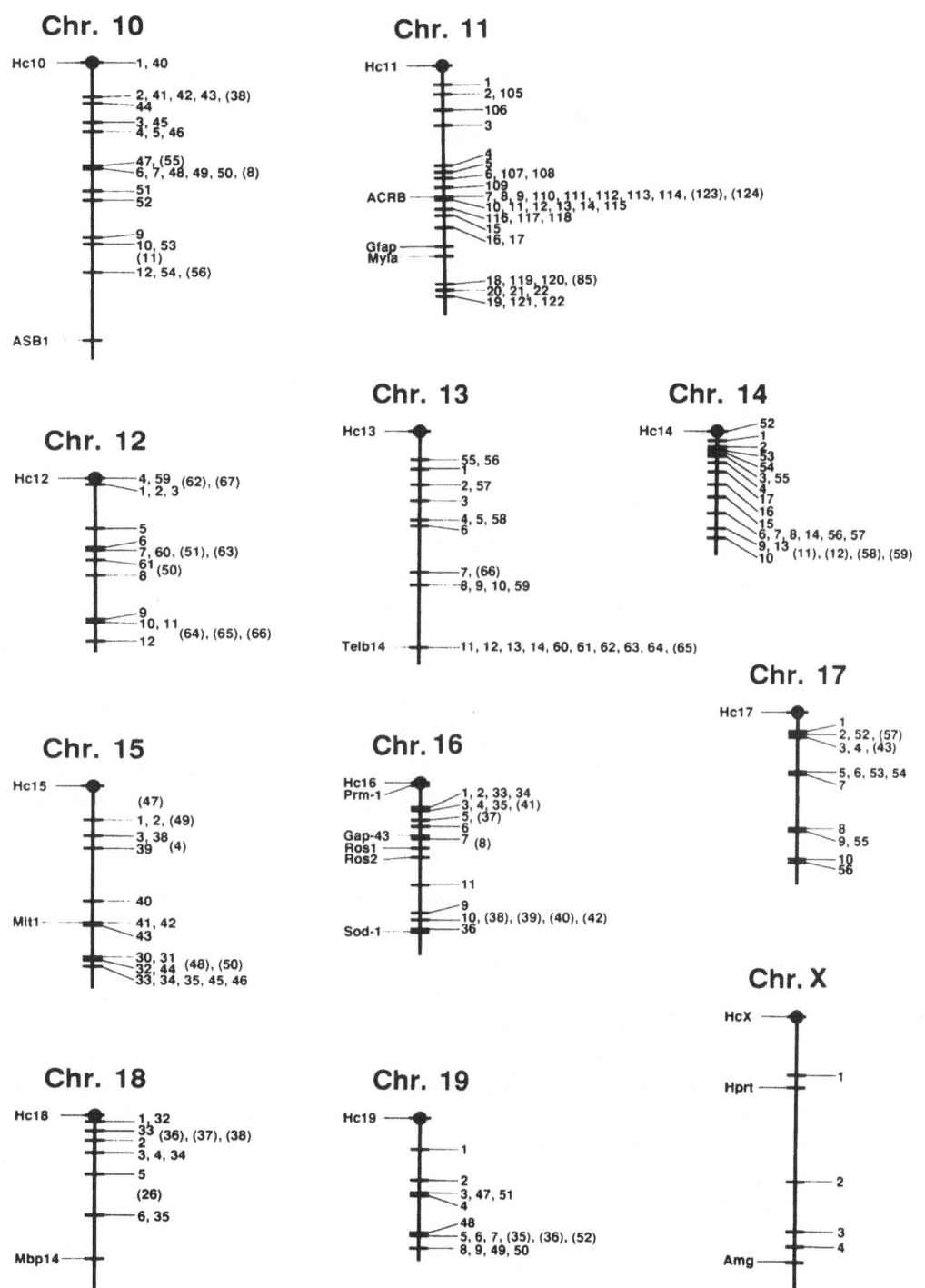

(Fig. 5.10 continued) text. Only the locus number is given without the chromosome and laboratory code is written on the *right side.* A 10-cM *bar* is shown on the *left side* of the chromosome 1 map [15]. (By permission of VCH)

different crosses and by identifying 29 B-specific loci that differed both between B6 and D2, and B6 and *M. spretus*. The 29 B-specific loci mapped to 16 chromosomes and additional 13 non-RLGS loci were used to align the remaining chromosomes with the RLGS loci [14]. In the case of the 73 reference loci that were used to anchor the HSH genetic maps [14], 33 of these were previously mapped in the BXD RI strains. The coincident anchor loci's relative positions were in the same order in all of the crosses and gave roughly similar overall positions on the different genetic maps [14].

5.6
Discussion

The RLGS analyses of congenic strains, RI strains, and backcrosses readily show that it is possible to identify extensive amounts of genetic variation among laboratory strains and furthermore, between laboratory mice and divergent *Mus* species. Moreover, it is possible to use conventional genetic crosses and other methods to localize these loci in the mouse genome and to produce ordered genetic maps that give the location of these loci on all of mouse chromosomes. In this regard, these analyses clearly demonstrate that RLGS is a powerful genomic scanning method able to sample loci from all of the complex mammalian genome. This demonstration is critical for applying RLGS methods to studies involving the loss of heterozygosity (LOH) in model tumor systems, for characterizing deletions and gene amplifications in tumorigenesis ([25] see Chap. 7), for analyzing changes in genomic methylation during development [26], and for characterizing imprinted loci across the mouse genome ([27], see Chap. 6).

The characterization of genetic variation in RLGS patterns has been demonstrated recently for human genomic DNA [28]. A series of spots that behaved as allelic variation were identified in the analysis of parent and progeny genomic DNAs. These spots segregated as codominant spots with a half-intensity in the parental DNAs, as a full-intensity spot at one allelic location, and as a missing spot in the alternative location. Electronic scanning of the autoradiograph and quantitative determination of spot intensity were the basis of these results. We determined quantitative variation in spot intensity and calculated coefficients of variation to determine whether full- and half-intensity labeling of spots could be distinguished. In our analysis of codominant *M. spretus*-specific spots, there was the advantage of comparing the segregating progeny with the original B6 and *M. spretus* parental profiles. From this, the parent-specific variation could be established. The segregation of spot intensities could be determined by comparing the labeling of non-variant spots in the same region. Hence, although our analyses lacked the precision of the electronic determination, it had the experimental advantage of closely associated control spots that allowed, therefore, a compensation for differences in isotope labeling between gels and other factors that might perturb the absolute levels of spot intensity between samples.

The demonstration that codominant loci can be mapped in backcross progeny is critically important because it increases the number of loci that can be identified and mapped in backcross analyses. This information's most immediate application comes from the use of RLGS methods with pooled samples of DNA for segregating progeny identified by a common recessive trait or complex trait phenotype (our unpubl. data). In this instance, a pool of backcross progeny will be homozygous for closely linked loci; alleles of the opposite parent of the F_1 hybrid will not be present. Furthermore, the intensity of the mutant parental strain spots associated with RLGS loci that are closely linked to the mutant will appear to be of full intensity in comparison to the levels observed for unlinked loci.

In principle, it is relatively easy to integrate the RLGS genetic maps with other genetic maps identified by methods such as the simple sequence repeat loci (Mit). To date, more than 13 000 loci in total have been characterized in the mouse [29]. Additionally, it is possible to analyze a large sample of these markers in the same BSS progeny. A partial integration of the RLGS genetic map with the Frederick Cancer Center functional map was accomplished by using the BSS map to integrate centromeric distances from functional genes included in the Frederick interspecific backcross [30]. In practice, however, the integration of these additional loci into the BSS backcross is relatively expensive and may not be a cost-effective strategy. Nevertheless, one may consider the analysis of specific genomic regions of special interest to various projects as a preliminary step to identifying candidate RLGS loci in mutant gene regions for various positional cloning experiments.

The use of *Not*I as the restriction landmark for these studies suggests a high probability that a significant number of these RLGS loci will be associated with CpG islands of functional genes. From this viewpoint, it is probable that RLGS loci in the mouse will be used to identify homologous loci in other species and it should therefore be possible to use these sites for comparative mapping analyses. The development of spot cloning techniques allows for the development of cloned probes that can be used for these analyses [31]. We have used a probe made up of the cloned sequences from the characterization of an imprinted mouse locus to analyze human somatic cell hybrids and also to conduct FISH analysis on human chromosome 7. We noted that the probe identified unique sequence fragments in Southern analysis of human genomic DNA and could observe a clear pattern of hybridization on metaphase chromosomes. Using these approaches, a map location on an evolutionarily conserved region of human 5q and mouse Chr 11 was identified ([11,27] see Chap. 6).

One of the major advantages of the RLGS method of analysis is that it does not rely upon cloned probes or sequences to apply it to a new species. Hence, it is possible to use RLGS to analyze genetic map-poor species [32] for the linkage of segregating traits or recessive mutations. If RLGS loci can be linked to a mutation or a trait, spot cloning can be used to recover these landmarks. Actually, this mapping method is very useful to analyze orphan genomes with mutant trait(s). Recently, we constructed a whole genetic map of Syrian hamster, using RLGS-spot mapping, and we also identified hamster cardiomyopathy locus on chromosome 9ga 2.1–b1, based on this genetic map [33,34]. These clones may then be used as probes for analysis in mouse crosses. This will indicate the comparative map position in the mouse genome and suggest candidate genes that map to the same region.

To summarize, the RLGS method has been utilized in mapping loci in congenic strains, RI strains, and interspecific backcrosses. The method is particularly well suited for identifying loci that segregate as dominant markers in a positive and negative fashion. However, it is also possible to recover a significant number of RLGS loci that differ quantitatively between individual backcrosses and segregate as codominant loci.

References

1. Lueders KK, Franke WN, Mietz JA, Kuff EL (1993) Genomic mapping of intracisternal A-particle proviral elements. Mammal Genome 4:69–77
2. Frankel WN, Coffin JM (1994) Endogenous nonecotropic proviruses mapped with oligonucleotide probes from the long terminal repeat region. Mammal Genome 5:275–281

3. Kozak C, Peters G, Pauley R, Morris V, Michalides R, Dudley J, Green M, Davisson M, Prakash O, Vaidya A, Hilgers J, Verstraeten A, Hynes N, Diggelmann H, Peterson D, Cohen JC, Dickson C, Sarkar N, Nusse R, Varmus H, Callahan R (1987) A standardized nomenclature for endogenous mouse mammary tumor viruses. J Virol 61:1651–1654

4. Siracusa LD, Jenkins NA, Copeland NG (1991) Identification and applications of repetitive probes for gene mapping in the mouse. Genetics 127:169–179

5. Serikawa T, Montagutelli X, Simon-Chazottes D, Guenet J-L (1992) Polymorphisms revealed by PCR with single, short-sized, arbitrary primers are reliable markers for mouse and rat gene mapping. Mammal Genome 3:65–72

6. Woodward SR, Sedweeks J, Teuscher C (1992) Random sequence oligonucleotide primers detect polymorphic DNA products which segregate in inbred strains of mice. Mammal Genome 3:73–78

7. Nadeau JH, Bedigian HG, Bouchard G, Denial T, Kosowsky M, Norberg R, Pugh S, Sargeant E, Turner R, Paigen B (1992) Multilocus markers for mouse genome analysis: PCR amplification based on single primers of arbitrary nucleotide sequence. Mammal Genome 3:55–64

8. Welsh J, Petersen C, McClelland M (1991) Polymorphisms generated by arbitrarily primed PCR in the mouse: application to strain identification and genetic mapping. Nucleic Acids Res 19:303–306

9. Postlethwait JH, Johnson SL, Midson CN, Taibot WS, Gates M, Ballinger EW, Africa D, Andrews R, Carl T, Eisen JS, Horne S, Kimmel CB, Hutchinson M, Johnson M, Rodriguez A (1994) A genetic linkage map for the zebrafish. Science 264:699–703

10. Birkenmeier EH, Schneider U, Thurston SJ (1992) Fingerprinting genomes by use of PCR with primers that encode protein motifs or contain sequences that regulate gene expression [published erratum appears in Mammal Genome 1993:4(2):133]. Mammal Genome 3:537–545

11. Kalcheva I, Plass C, Sait S, Eddy R, Shows T, Watkins-Chow D, Camper S, Shibata H, Ueda T, Takagi N, Hayashizaki Y, Chapman V (1995) Comparative mapping of the imprinted U2afbpL gene on mouse chromosome 11 and human chromosome 5. Cytogenet. Cell Genet 68:19–24

12. Imoto H, Hirotsune S, Muramatsu M, Okuda K, Sugimoto O, Chapman VM, Hayashizaki Y (1994) Direct determination of NotI cleavage sites in the genomic DNA of adult mouse kidney and human trophoblast using whole-range restriction landmark genomic scanning. DNA Res 1:239–243

13. Shibata H, Hirotsune S, Okazaki Y, Komatsubara H, Muramatsu M, Takagi N, Ueda T, Shiroishi T, Moriwaki K, Katsuki M, Chapman VM, Hayashizaki Y (1994) Genetic mapping and systematic screening of mouse endogenously imprinted loci detected with restriction landmark genome scanning method (RLGS). Mammal Genome 5:797–800

14. Hayashizaki Y, Hirotsune S, Okazaki Y, Shibata H, Muramatsu M, Kawai J, Hirasawa T, Shiroishi T, Watanabe S, Moriwaki K, Taylor B, Matsuda Y, Elliott R, Manly K, Chapman VM (1994) A genetic linkage map of the mouse using Restriction Landmark Genomic Scanning (RLGS). Genetics 138:1207–1238

15. Okuizumi H, Okazaki Y, Ohsumi T, Hanami T, Mizuno Y, Muramatsu M, Hayashizaki Y, Plass C, Chapman VM (1995) A single gel analysis of 575 dominant and codominant restriction landmark genomic scanning loci in mice interspecific backcross progeny. Electrophoresis 16:253–260

16. Okazaki Y, Hirose K, Hirotsune S, Okuizumi H, Sasaki N, Ohsumi T, Yoshiki A, Kusakabe M, Muramatsu M, Kawai J, Watanabe S, Plass C, Chapman VM, Nakao K, Katsuki M, Hayashizaki Y (1995) Direct detection and isolation of RLGS spot DNA marker tightly linked to a specific trait using RLGS spot-bombing method. Proc Natl Acad Sci USA 92:5610–5614

17. Okazaki Y, Okuizumi H, Sasaki N, Ohsumi T, Kuromitsu J, Kataoka H, Muramatsu M, Iwadate A, Hirota N, Kitajima M, Plass C, Chapman VM, Hayashizaki Y (1994) A genetic linkage map of the mouse using an expanded production system of restriction landmark genomic scanning (RLGS Ver. 1.8). Biochem Biophys Res Commun 205:1922–1929

18. Lander ES, Schork NJ (1994) Genetic dissection of complex traits. Science 265:2037–2048

19. Matsuda Y, Manly KF, Chapman VM (1993) In situ analysis of centromere segregation in C57BL/6 × Mus spretus interspecific backcrosses. Mammal Genome 4:475–480

20. Manly KF (1993) A macintosh program for strage and analysis of experimental genetic mapping data. Mammal Genome 4:303–313

21. Held WA, Pazik J, O'Brien JG, Kerns K, Gobey M, Meis R, Kenney L, Rustum Y (1994) Genetic analysis of liver tumorigenesis in SV40 T antigen transgenic mice implies a role for imprinted genes. Cancer Res 54:6489–6495

22. Ohsumi T, Okazaki Y, Okuizumi H, Shibata K, Hanami T, Mizuno Y, Takahara T, Sasaki N, Ueda M, Muramatsu M, Kerns KA, Chapman VM, Held WA, Hayashizaki Y (1995) Loss of hetrozygosity in chromosome 1, 5, 7 and 13 in mouse hepatoma detected by systematic genome-wide scanning using RLGS genetic map. Biochem Biophys Res Commun 212:632–639

23. Dietrich WF, Radany EH, Smigh JS, Bishop JM, Hahahan D, Lander ES (1994) Genome-wide search for loss of heterozygosity in transgenic mouse tumors reveals candidate tumor suppressor genes on chromosomes 9 and 16. Proc Natl Acad Sci USA 91:9451–9455

24. Hirotsune S, Takahara T, Sasaki N, Hirose K, Yoshiki A, Ohashi A, Kusakabe M, Murakami Y, Muramatsu M, Watanabe S, Nakao K, Katsuki M, Hayashizaki Y (1995) The reeler gene encodes a protein with an EGF-like motif expressed by pioneer neurons. Nat Genet 10:77–83

25. Hirotsune S, Hatada I, Komatsubara H, Nagai H, Kuma K, Kobayakawa K, Kawara T, Nakagawara A, Fujii K, Mukai T, Hayashizaki Y (1992) New approach for detection of amplification in cancer DNA using Restriction Landmark Genomic Scanning. Cancer Res 52:3642–3647

26. Kawai J, Hirotsune S, Hirose K, Fushiki S, Watanabe S, Hayashizaki Y (1993) Methylation profiles of genomic DNA of mouse developmental brain detected by restriction landmark genomic scanning (RLGS) method. Nucleic Acids Res 21:5604–5608

27. Hayashizaki Y, Shibata H, Hirotsune S, Sugino H, Okazaki Y, Sasaki N, Hirose K, Imoto H, Okuizumi H, Muramatsu M, Komatsubara H, Shiroishi T, Moriwaki K, Katsuki M, Hatano N, Sasaki H, Ueda T, Mise N, Takagi N, Plass C, Chapman VM (1994) Identification of an imprinted U2af binding protein related sequence on mouse chromosome 11 using the RLGS method. Nat Genet 6:33–40

28. Asakawa J-I, Kuick R, Nell JV, Kodaira M, Satoh C, Hanash SM (1994) Genetic variation detected by quantitative analysis of end-labeled genomic DNA fragments. Proc Natl Acad Sci USA 91:9052–9056

29. Chromosome Committee (1996) Master locus list. Mammal Genome 6:S2–S27

30. Ceci JD, Matsuda Y, Grubber JM, Jenkins NA, Copeland NG, Chapman VM (1994) Interspecific backcrosses provide an important new tool for centromere mapping of mouse chromosomes. Genomics 19:515–524

31. Hirotsune S, Shibata H, Okazaki Y, Sugino H, Imoto H, Sasaki N, Okuizumi H, Muramatsu M, Plass C, Chapman VM, Tamatsukuri S, Miyamoto C, Furuichi Y, Hayashizaki Y (1993) Molecular cloning of polymorphic markers on RLGS gel using the spot target cloning method. Biochem Biophys Res Commun 194:1406–1412

32. Jacob H (1996) A landmark for orphan genomes? Nat Genet 13:14–15

33. Okazaki Y, Okuizumi H, Ohsumi T, Nomura O, Takada S, Kamiya M, Sasaki N, Matsuda Y, Nishimura M, Tagaya O, Muramatsu M, Hayashizaki Y (1996) A genetic linlage map of the Syrian hamster and localization of cardiomyopathy locus on chromosome 9qa2. 1-b1 using RLGS spot-mapping. Nat Genet 13:87–90

34. Okuizumi H, Ohsumi T, Sasaki N, Imoto H, Mizuno Y, Hanami T, Yamashita H, Kamiya M, Takada S, Kitamura A, Muramatsu M, Nishimura M, Mori M, Matsuda Y, Tagaya O, Okazaki Y, Hayashizaki Y (1997) Linkage map of Syrian hamster with restriction landmark genomic scanning. Mammal Genome 8 (in press)

35. Okuizumi H, Okazaki Y, Ohsumi T, Hayashizaki Y, Plass C, Chapman VM (1995) Genetic mapping of restriction landmark genomic scanning loci in the mouse..Electrophoresis 16:233–240

Chapter 6

Application of RLGS to Screening Endogeneously Imprinted Genes

HIDEO SHIBATA and CHRISTOPH PLASS

Contents

6.1
Scanning Methylation State with RLGS Using Methylation-Sensitive Endonuclease (RLGS-M)

Landmark restriction sites on the whole genome can be visualized and screened with the RLGS technique. If recognition sites of methylation-sensitive endonuclease are used for landmark sites, the appearance of RLGS spots depends not only the interspecific polymorphism but on the methylation state of landmark sites. When a landmark site was methylated and resistant to digestion, the corresponding RLGS spot disappeared and vice versa. Therefore the RLGS technique is applicable to scanning the methylation state of landmark sites on the whole genome when using methylation-sensitive endonuclease for landmark sites. We have named this technique restriction landmark genome scanning using methylation-sensitive endonuclease (RLGS-M), and have used it to screen endogenously imprinted loci/genes [1,2], genes expressing with a developmental stage specificity [3], and cell type specificity [4].

In RLGS-M analysis, the spot intensity reflects the copy number of each landmark site, because the number of endlabeled molecules at landmark cleavage sites corresponds with the incorporated radioactivity (Fig. 6.1). Uniform levels of spot intensity are observed in diploid tissues; this is consistent with the cleavage and endlabeling of both alleles for most landmark sites. However, when methylation-sensitive enzymes are used for the landmark cleavage sites, the spots show half-intensity or zero intensity,

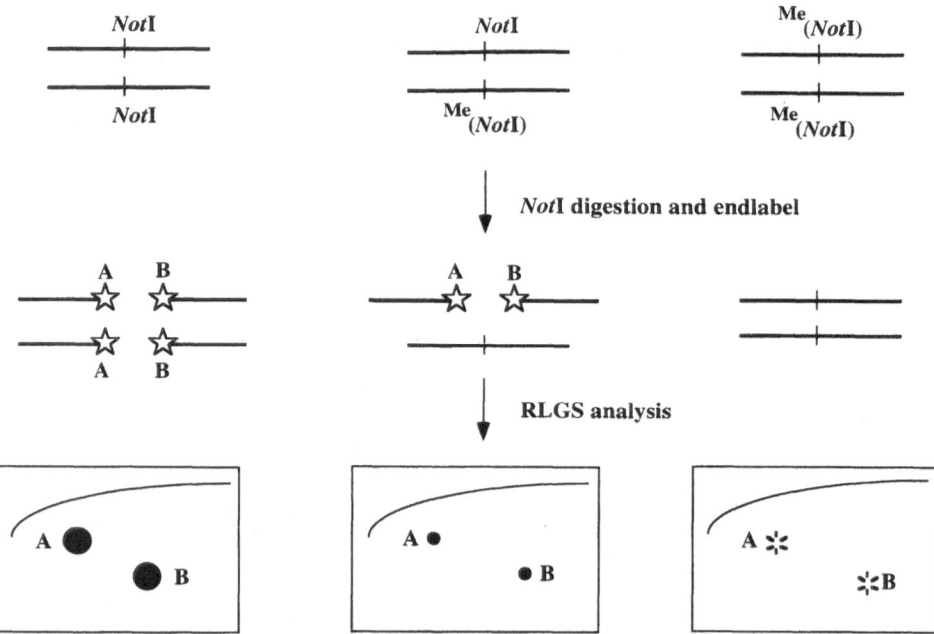

Fig. 6.1. Methylation state of landmark *Not*I site and intensity of RLGS spot. Diploid genomic DNA shown in a *pair of lines*. *Not*I sites or methylated *Not*I sites shown in *vertical bars*. Landmark sites radiolabeled with [α-^{32} P]-dCTP and [α^{32} P]-dGTP symbolized with *star marks*. In schemed RLGS profile, *large and small circles* show RLGS spots of diploid intensity and haploid intensity, respectively. *Dark and shaded circles* show RLGs spots from the left- and right-side fragment of the same landmark *Not*I site. *Left column* indicates a case in which *Not*I sites are unmethylated on both alleles; *middle column* methylated on one of alleles; *right column* methylated on both alleles

respectively, when one or both of the alleles is methylated (Fig. 6.1). GC-rich enzymes such as *Not*I (GCGGCCGC) or *Bss*HII (GCGCGC) cleave sequences that are preferentially located on CpG islands [5]. Approximately 89% of *Not*I sites and 74% of *Bss*HII sites were estimated to be in CpG islands [5]. Thus, the methylation status of CpG islands could be evaluated using *Not*I or *Bss*HII as landmarks. Moreover, since methylation has been correlated with the allele-specific expression of imprinted genes, it has been possible to screen *Not*I and *Bss*HII sites for allele-specific patterns of methylation that are parent-specific and candidates for imprinted loci.

6.2
Introduction of Genomic Imprinting

6.2.1
Genomic Imprinting

Genomic imprinting, in a narrow definition, describes the phenomenon when a gene does not follow the Mendelian rule of expressing only one parentally specific allele. As one application of the RLGS technique, we screened imprinted loci throughout the whole mouse genome to identify novel genes [1,2]. In this chapter we describe this screening technique and also the characteristics of identified novel imprinted genes.

Genomic imprinting in mammals results in genes or chromosomes that retain some memory of their gametic origin and subsequently alter patterns of cytosine methylation and expression of alleles in a parent-specific manner. The phenomenon of imprinting was initially described by Crouse (1960) on Sciara, in which the paternally derived X chromosome was selectively deleted [6]. Imprinting phenomena were reported also in mealbugs and marsupials [7,8]. In mammals, the phenomena of imprinting was first reported for X chromosome inactivation of mouse extraembryonic tissues in which the paternally transmitted X chromosome was selectively inactivated [9].

Experimentally produced gynogenotes and androgenotes fail to develop beyond the early to mid-gestational period of embryogenesis, with characteristic underdevelopment of the embryonic linages in androgenotes, and a deficiency of extraembryonic growth in gynogenotes [10–12]. Similar parental-specific effects were also suggested by the general failure of parthenogenesis in mammals.

Regions of parental-specific effects have been identified using mouse T^{hp} mutant, human disease, or uniparental duplication/ deletion disomic mice [13–15]. These results strongly suggest that gametes of both parents are essential for normal mammalian development and that there is gamete-specific modification of specific genes.

6.2.2
Reported Imprinted Genes

In 1991 the first imprinted gene, Igf2r, was identified in the imprinted region, Tme locus of T^{hp} deletion mutant [14]. Subsequently, approximately 20 imprinted genes were identified and characterized in humans and mice (Table 6.1). Human homologs for mouse imprinted genes were also imprinted for *Igf2*, *H19*, *Snrpn*, *p57*^{KIP2}, but not for *Igf2r* and *U2afbp-rs* (Table 6.1). In human diseases, genomic imprinting has been implicated in the Beckwith-Wiedeman syndrome, the Angelman syndrome, the Prader-Willi syndrome, and in the onset of cancers [16–18].

Table 6.1. Imprinted genes in mouse and human

	Chromosomal location	Expressed allele	Human homolog	Reference
Gnas	Distal chr2	?	–	13
Peg1/Mest	Proximal chr6	Paternal	–	40
Apoc2	Proximal chr7	Paternal	–	40
Snrpn	Central chr7	Paternal	Imprinted (15q11–13)	15, 41, 42
Znf127	Central chr7	?	–	13
DN34	Central chr7	?	–	13
IPW	–	Paternal	Imprinted (15q11–13)	43
PAR-1	–	Paternal	Imprinted (15q11–13)	44
PAR-5	–	Paternal	Imprinted (15q11–13)	44
Ins1, 2	Distal chr7	Paternal	Not imprinted (11p15)	45
H19	Distal chr7	Maternal	Imprinted (11p15)	46, 47
Igf2	Distal chr7	Paternal	Imprinted (11p15)	48, 49
Mash2	Distal chr7	Paternal	–	50
p57KIP2	Distal chr7	Maternal	Imprinted (11p15)	51
cdc25Mm	Central chr9	Paternal	–	35
U2afbp-rs	Proximal chr11	Paternal	Not imprinted (5q31)	34, 38
Igf2r	Proximal chr17	Maternal	Not imprinted (6q26)	14, 52
Mas	Proximal chr17	Paternal	–	53
Xist	chrX	Paternal	–	54

6.2.3
Methylation and Imprinted Gene

Epigenetic modifications of the genome in the form of methylation of the 5′ position of cytosine is the most compelling mechanism for imprinting. Allele-specific methylation has been identified on *Igf2*, *Igf2r*, *H19*, *Snrpn*, *U2afbp-rs*, and *Xist* [18–23]. *Igf2*, *Snrpn*, *U2afbp-rs*, and *Xist* were expressed from the paternally transmitted allele, and *Igf2r* and *H19* from the maternally transmitted allele. Regions hypermethylated on the nonexpressed allele and hypomethylated on the expressed allele were identified in *Igf2r*, *H19*, *Snrpn*, *U2afbp-rs*, and *Xist* [20–24]. Conversely, regions hypermethylated on the expressed allele and hypomethylated on the nonexpressed allele were identified in *Igf2* and *Igf2r* [19,20]. Thus, allele-specific methylation appears to correspond with the imprinted expression of these genes. Imprinted patterns of expression were established at the implantation stages shown using androgenotes and gynecogenotes [25,26].

Additional studies demonstrate, with the use of DNA methyltransferase-deficient mice, the importance of DNA methylation as a mechanism for regulating imprinted patterns of expression [27,28]. *H19* was expressed biallelically and *Igf2* and *Igf2r* were silent in methyltransferase-deficient mice [28].

It is important to identify additional imprinted genes in order to understand the mechanisms controlling genomic imprinting phenomena and the role of imprinted regulation in mammalian development. The total number of imprinted genes in the mouse genome has been estimated to be as high as 100 [1,2,29]. The initial attempts to identify chromosomal regions that carry imprinted loci have used translocation duplication/deletion uniparental disomic mice [13]. Several imprinted regions of the mouse genome have been identified with this approach, but there are several inherent limitations with this strategy. First, altered phenotypes may not be detectable in all cases, even if imprinted genes are located in the disomic region. Second, mice with Robertsonian or reciprocal translocations may not cover the entire mouse genome. Third, the chromosomal regions tested in these duplication/deletion tests are fairly extensive and it is difficult to limit the studies to defined gene regions.

Several transgenes showing imprinted patterns of methylation have been identified. However, recent work on imprinted transgenes suggests that the combination of integrated DNA and flanking genomic DNA determines the imprinted phenotype and that imprinted DNA methylation does not extend to flanking genomic DNA [30].

6.3
Strategy of Screening of Imprinted Loci

6.3.1
Allele-Specific Methylation in Imprinted Genes

If we assume that DNA methylation is a compelling mechanism for genomic imprinting, then it follows that the identification of endogenous loci showing parental allele-specific methylation provides an effective strategy for finding imprinted genes. RLGS or RLGS-M is an efficient method for screening the genome for imprinted loci or genes, since it is a multiplex method that can detect the methylation status of a several thousand loci in one gel using methylation-sensitive enzymes.

6.3.2
Identifying Loci of Imprinted Methylation with RLGS

Candidate RLGS spots for imprinted loci were identified using several criteria: (1) RLGS spots that were variant between parental strains were identified in order to follow the transmission of spots in reciprocal hybrid progeny; (2) spots that showed half-intensity to nonpolymorphic spots in the parental RLGS profile indicated that methylated landmark sites on one allele were not cleaved; (3) strain-specific spots that appeared in the profile of the F_1 between two different strains and disappeared in profile for its reciprocal cross.

An RLGS spot showing imprinted transmission is indicated in Fig. 6.2, which shows the transmission pattern of RLGS spot *Irlgs3* in reciprocal crosses between C57BL/6J (B6) and DBA/2J (D2). RLGS spot *Irlgs3* is a polymorphic *Not*I landmark between B6 and D2, specific for D2. It shows a half strength in spot intensity compared with other neighboring spots, suggesting the *Not*I site for *Irlgs3* is methylated on one allele and demethylated on the other. *Irlgs3* is not seen in the RLGS profile for F_1 hybrid (B6 × D2)F_1 (Fig. 6.2a); however, it was detected in the profile for (D2 × B6)F_1 (Fig. 6.2b). This transmission pattern of *Irlgs3* suggests that the *Not*I site

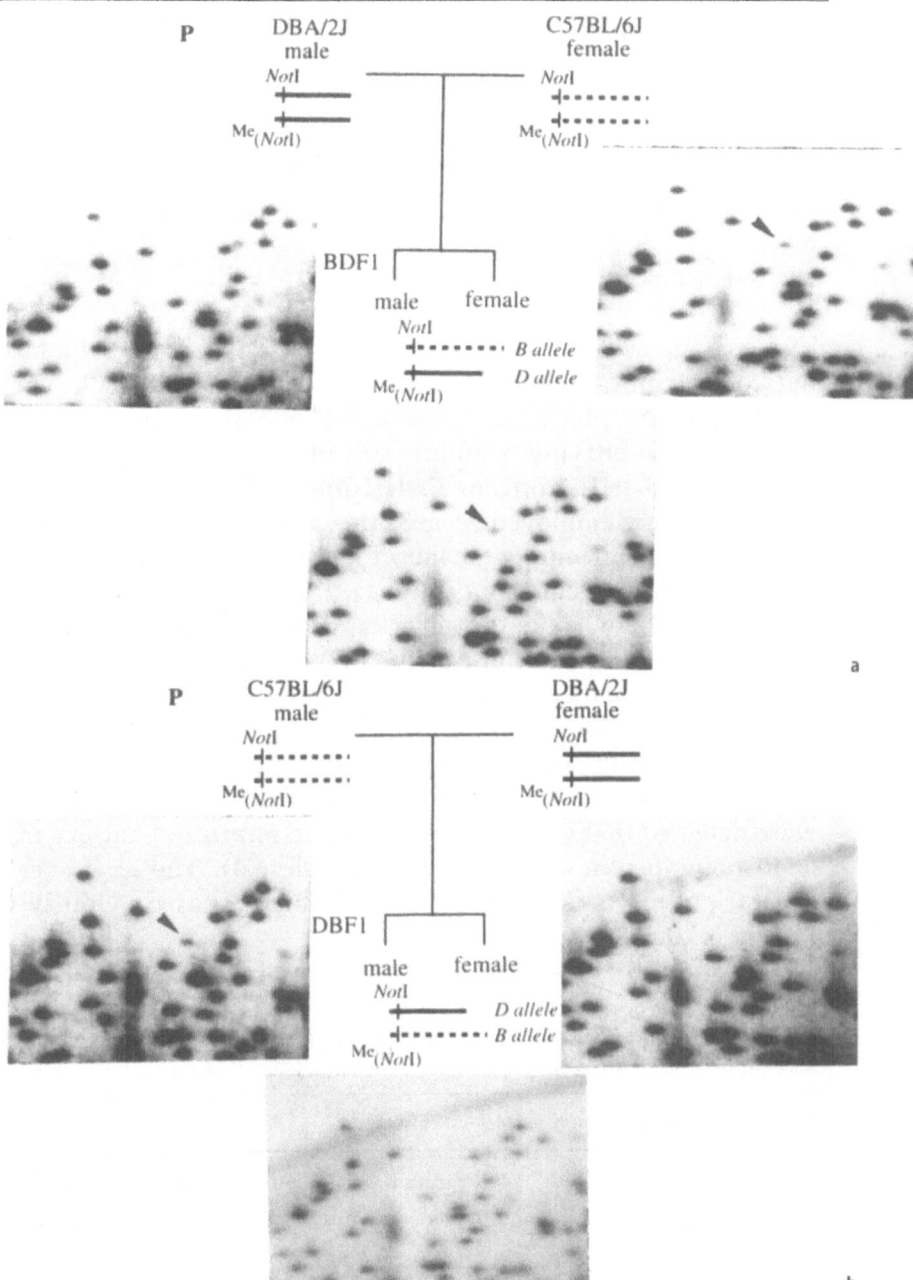

Fig. 6.2a,b. The transmission of *Irlgs3* in reciprocal crosses of (C57BL/6J female × DBA/2J male)F₁ (**a**) and (DBA/2J female × C57BL/6J male)F₁ (**b**). Restriction enzymes of *Not*I, *Eco*RV, and *Mbo*I were used for landmark (E_L), E_B, and E_C, respectively. *Arrowheads* indicate the spot *Irlgs3*. *Not*I and ^Me(*Not*I) indicate demethylated or methylated *Not*I landmark. C57BL/6J (B) allele and DBA/2J (D) allele are shown in *dotted* and *normal line*, respectively

for *Irlgs3* is demethylated on the maternally transmitted allele and methylated on the paternally transmitted allele.

6.4
Screening of Imprinted Loci Using RLGS-M

6.4.1
Scale of Screening and Identified Loci

The enzyme combinations used for RLGS analysis and the numbers of polymorphic loci screened are shown in Table 6.2. A total of 3040 polymorphic RLGS spots of *Not*I or *Bss*HII landmarks were screened with enzyme combinations of *Not*I-*Eco*RV-*Mbo*I, *Not*I-*Sph*I-*Hin*fI, *Bss*HII-*Hin*fI, or *Bss*HII-*Mbo*I. The laboratory strains B6 and D2 were employed because they showed the greatest degree of polymorphism among the laboratory mice, 13.4% in RLGS profile. B6- or D2-specific spots appeared in RLGS profiles for BDF$_1$ and DBF$_1$. A few spots, however, appeared in only one of both reciprocal matings, but not in the other. Three *Not*I spots and a *Bss*HII spot were detected in the mating between B6 and D2, whose transmission followed genomic imprinting. Further screening was performed using matings between B6 and divergent strain of *Mus musculus* (PWK) or *Mus molossinus* (MSM). Eight spots were detected that were transmitted in an imprinted pattern from 3040 polymorphic spots screened (Table 6.2). The character of these eight spots is summarized in Table 6.3, as previously reported [1,2].

Table 6.2. Summary of RLGS landmarks and numbers of spots employed for the screening of imprinted spots [1]

Landmark enzymes	*Not*I		*Bss*HII	
Combination of enzymes	*Eco*RV-*Mbo*I	*Sph*I-*Hin*fI	*Hin*fI	*Mbo*I
Total spots in profile	2300	2300	1150	1300
Polymorphic spots(%)				
B6 and D2	310 (13.4)	310 (13.4)	160 (13.4)	170 (13.4)
B6 and PWK	720 (31.3)	–	–	410 (31.3)
B6 and MSM	610 (26.7)	–	–	350 (26.7)
total				
	1950		1090	
Imprinted loci	7		1	

Table 6.3. Summary of endogeneoulsy imprinted loci identified from the screening with RLGS-M [1]

	Spot number							
	1	2	3	4[e]	5	6[f]	7	8
Strain[a]	D	D M P	B	D	P	M P	M P	D P
Restriction landmark[b]	N	N	N	Bs	N	N	N	N
Copy number	1	1	1	2,4	1	1	1	1
Combination of mating[c]								
B6 × DBA2								
extraembryonic	−	+	+	(4)	np	np	np	nd
embryonic	−	+	+	(4)	np	np	np	nd
DBA2 × B6								
extraembryonic	+	−	−	(2)	np	np	np	nd
embryonic	+	−	−	(2)	np	np	np	nd
B6 × PWK	np	+	+	np	−	−	+	−
PWK × B6	np	−	−	np	+	+	−	+
B6 × MSM	np	+	np	np	np	+	+	np
MSM × B6	np	−	np	np	np	+	−	np
Paternal or maternal transmission[d]	m	p	m	p	m	m	p	m
Chromosomal location	9	11	9					
Imprinted locus name	Irlgs1	U2afbprs	Irlgs3					
RLGS spot name	CD(B)700	CD(B)461	CB(D)401					
Locus name	D9Ncvs58	D11Ncvs75	D9Ncvs53					

[a] The inbred strain whose spot is strain-specific, D; DBA/2J, B; C57BL/6J, P; PWK (an inbred line derived from Mus musculus musculus), M; MSM (an inbred line derived from Mus musculus molossinus).
[b] N; NotI, Bs; BssHII.
[c] np: not polymorphic; nd: not determined; + appearance of spot; − disappearance of spot.
[d] p: spot appeared in the paternal transmission; m: spot appeared in the maternal transmission.
[e] The numbers in parentheses represent the copy number of spot 4 in independent samples which were analyzed by RLGS-M.
[f] The transmission of spot 8 follows the imprinting rule in the cross between B6 and PWK, but not in that between B6 and MSM.

6.4.2
Confirmation of Imprinted Transmission Using Backcross Pedigree

The imprinted transmission of these spots can be confirmed using a backcross mating of B6 and D2. Figure 6.3 shows the transmission pattern of RLGS spot *Irlgs2*. *Irlgs2*, a D2-specific *NotI* landmark, disappeared in the RLGS profile of DBF₁ when it was transmitted through the maternal gamete (Fig. 6.3b). *Irlgs2* reappeared through paternal transmission in the following generation of backcross 1 (BC1). However, *Irlgs2* was not detected in BC1 through maternal transmission. The imprinted transmission pat-

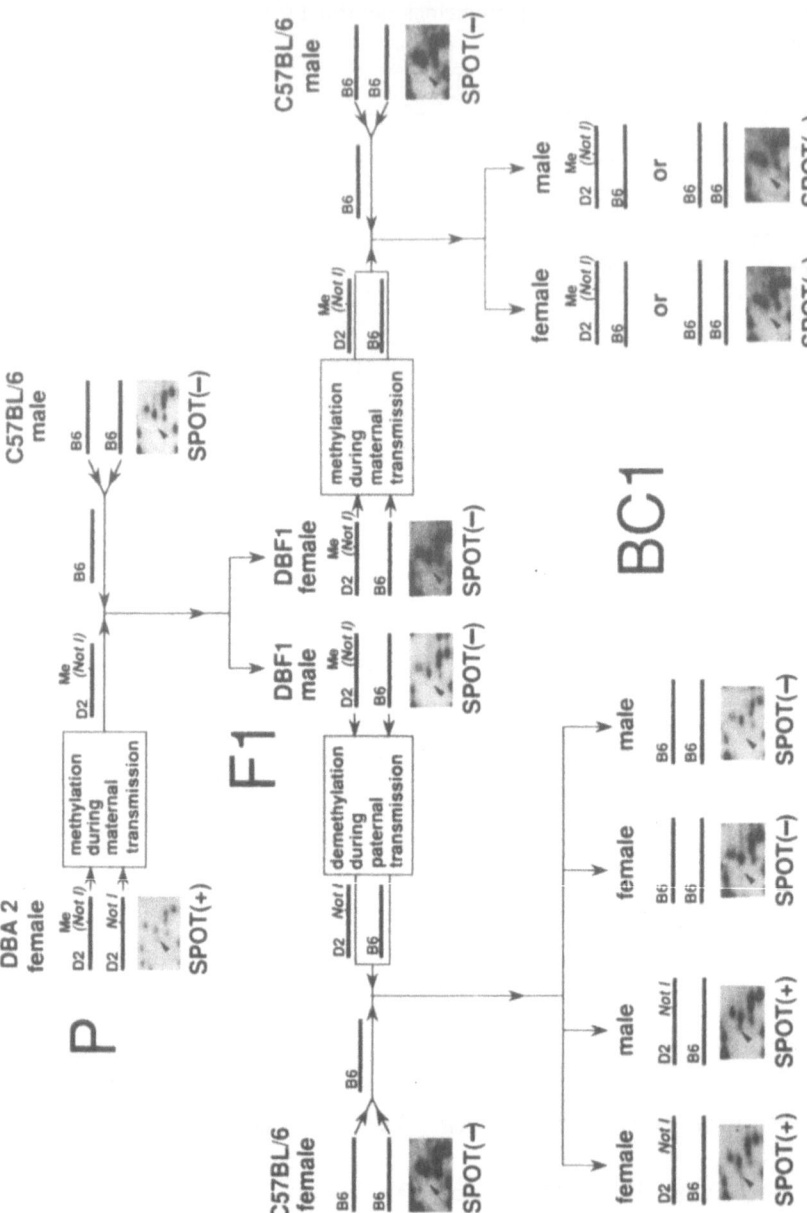

Fig. 6.3a,b. Transmission of imprinted spot *Irlgs2* using matings between C57BL/6 and DBA/2. Two lines originating from reciprocal crosses B6 female × D2 male (**a**) and D2 female × B6 male (**b**) were used. Restriction enzymes of *NotI*, *EcoRV*, and *MboI* were used for landmark (E$_L$), E$_B$, and E$_O$ respectively. *Arrowhead* indicates *Irlgs2*. *Solid bars* indicate alleles on *Irlgs2/U2afbp-rs* locus. *NotI* and ^{Me}(*NotI*) indicate demethylated or methylated *NotI* landmark. Abbreviations: *P* progenitor; *BC1* backcross 1. [34] (By permission of Nature Genetics)

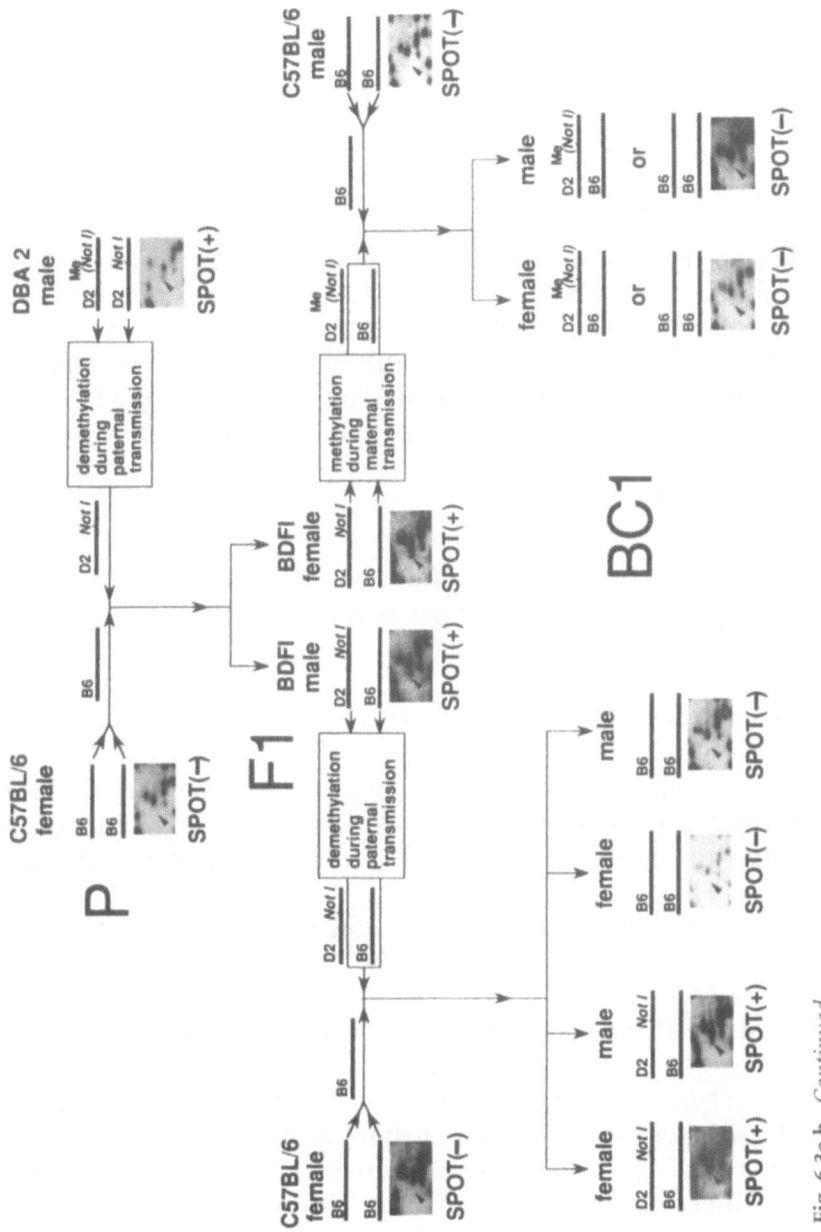

Fig. 6.3a,b. *Continued*

tern of *Irlgs2* was confirmed using reciprocal matings as shown in Fig. 6.3a.

Methylation-sensitive enzyme sites that have GC-rich recognition sequences are located preferentially in CpG islands. Approximately 89% of *Not*I sites and 74% of *Bss*HII sites were estimated to be in CpG islands [5]. There are an estimated 37 000 CpG islands in the mouse genome [31]. Approximately 2400 *Not*I loci were identified in the mouse genome from an expanded analysis of RLGS profiles of *Not*I landmarks [32]. Using these estimations, about 40% of *Not*I sites and about 3.5% of CpG islands were screened for their methylation state. The identification of eight imprinted loci among the 3040 *Not*I and *Bss*HII landmarks examined suggests that there might be approximately 100 imprinted loci in the entire genome.

6.5
Confirmation of the Parental Allele-Specific Methylation of *Not*I Sites for Imprinting Spot *Irlgs2*

6.5.1
Cloning of Genomic Fragments for *Irlgs2*

For the molecular cloning of genomic fragments corresponding to *Irlgs2*, a restriction trapper was used [33]. Precise procedures using restriction trappers are described in Chapter 4. A 0.8-kb *Not*I-*Mbo*I genomic fragment of D2 for *Irlgs2* was cloned into pBlueScriptII KS+ (Stratagene, Carifornia).

6.5.2
Parental Allele-Specific Methylation of the *Not*I Site of *Irlgs2*

Parental allele-specific methylation of the *Not*I site was confirmed with Southern hybridization analysis using cloned genomic fragments for *Irlgs2* and *Irlgs3*, respectively. In Fig. 6.4, B6 genomic DNA digested with *Eco*RV gives a hybridization signal in more than 20 kb (lane 2). A double digest with *Eco*RV and *Not*I shows two bands of *Eco*RV-*Eco*RV fragments and a 1.8-kb *Not*I-*Eco*RV fragment (lane 3), suggesting that the *Not*I site for *Irlgs2* is unmethylated in one allele and methylated in the other. The *Not*I-*Eco*RV fragment shows a fragment length polymorphism between B6 and D2 (lanes 3 and 7). Using this fragment length polymorphism, the parental allele can be distinguished in reciprocal crosses between B6 and D2. A hybridization signal for the 1.8-kb *NotI-Eco*RV fragment is derived from the B6 allele in BDF$_1$ (lanes 11 and 12) and from the D2 allele in DBF$_1$ (lanes 13 and 14), indicating that the *Irlgs2 NotI* site is unmethylated on the paternally derived allele and methylated on the maternally derived allele. The maternal allele-specific methylation of the *Irlgs2 NotI* site confirms the paternal transmission pattern of the spot in the RLGS profile.

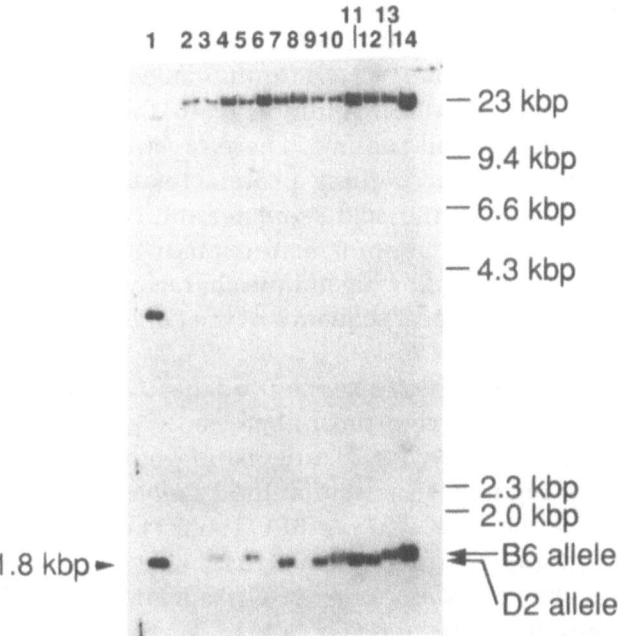

Fig. 6.4. The methylation pattern of the *Irlgs2 NotI* site following maternal imprint with Southern hybridization analysis. Five μg of genomic DNA was digested with restriction enzyme combination, electrophoresed in 30 cm 0.8% agarose gel, transferred to nylon membrane (Biodyne B, Pall Corp., Glenn Cove) and hybridized with radiorabeled 0.8-kb *NotI-Mbo*I genomic fragment for *Irlgs2*. *Lane 1* Genomic clone used for probe; *lane 2* male B6 genomic DNA digested with *Eco*RV; *lane 3* male B6 digested with *Eco*RV and *Not*I; *lane 4* female B6 digested with *Eco*RV; *lane 5* female B6 digested with *Eco*RV and *Not*I; *lane 6* male D2 digested with *Eco*RV; *lane 7* male D2 digested with *Eco*RV and *Not*I; *lane 8* female D2 digested with *Eco*RV; *lane 9* female D2 digested with *Eco*RV and *Not*I; *lane 10* female B6 and female D2 digested with *Eco*RV and *Not*I; *lane 11* male BDF$_1$ digested with *Eco*RV and *Not*I; *lane 12* female BDF$_1$ digested with *Eco*RV and *Not*I; *lane 13* male DBF$_1$ digested with *Eco*RV and *Not*I; *lane 14* female DBF$_1$ digested with *Eco*RV and *Not*I [34]. (By permission of Nature Genetics)

The *Irlgs3 Not*I site was also methylated specifically in the paternal allele after cloning the corresponding genomic fragment, which was consistent with the maternal transmission of *Irlgs3* in the RLGS profile (pers. comm., [35]).

6.6
Paternal Allele-Specific Expression of *U2afbp-rs*

Using a genomic clone of 0.8-kb *Not*I-*Mbo*I fragment, a transcript of approximately 2.9 kb was identified with Northern analysis [34]. cDNA of about 2.8 kb was isolated from the screening of the PWK liver cDNA library. The isolated cDNA carries an open reading frame coding a putative protein of 428 amino acid (51 kDa) and was a novel gene with significant homology to the human U2 auxiliary factor 35-kDa small subunit. Therefore we named this gene the U2 auxiliary factor binding protein related sequence, *U2afbp-rs*. The deduced amino acid sequence indicated that the protein has an RS domain, a common feature among nucleosomal proteins. This gene has the following unique characteristics: (1) no introns, and (2) a tandem repeat sequence of the *Fok*I family in the CpG island.

We examined allele-specific expression of the *U2afbp-rs* gene using a fragment length polymorphism between B6 and PWK in the region upstream to ORF, for distinguishing both parentally transmitted alleles. Figure 6.5 shows that the *U2afbp-rs* transcript was expressed from the PWK allele in BPF_1 (lane 3) and from the B6 allele in PBF_1 (lane 4) in reciprocal crosses between B6 and PWK, indicating that *U2afbp-rs* is a novel imprinted gene with paternal allele-specific expression (Fig. 6.5).

For *Irlgs3* the cdc25Mm gene was shown to be expressed extensively from the paternal allele, which was located at approximately 30 kb from the *Irlgs3 Not*I site (pers. comm., [35]).

6.7
Discussion

We screened endogenously imprinted loci throughout a whole mouse genome as one of the applications of the RLGS technique [1,2]. Eight candidate loci were identified from the screening (Tables 6.2, 6.3). Moreover, we expanded our screening to mouse imprinted loci, and a total of 13 candidates were identified

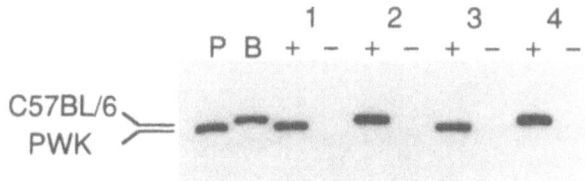

Fig. 6.5. Paternal allele-specific expression of mouse *U2afbp-rs*. *P* and *B* indicate genomic PCR for PWK and C57BL/6J (B6); + and − indicate with or without reverse transcriptase treatment. *Lane 1* 0.1 μg total RNA from PWK liver; *lane 2* 0.1 μg total RNA from C57BL/6J liver; *lane 3* 0.1 μg total RNA from (B6 × PWK)F₁ liver; *lane 4* 0.1 μg total RNA from (PWK × B6)F₁ liver. Please refer to [34] for information on primers and PCR conditions

(unpubl. data). We isolated genomic DNA fragments for two RLGS spots of these eight candidates and identified two imprinted genes, *cdc25Mm* and *U2afbp-rs*, the latter being a novel gene. These results suggest that our screening strategy works efficiently to identify endogeously imprinted loci and neighboring imprinted genes. Imprinted genes were found clustered in the *Snrpn* region on mouse chromosome 7 and also in the *H19-Igf2* region on mouse chromosome 7. The regions of imprinted regulation were thought to span more than several hundred kb. Our screening of imprinted loci could detect sites of imprinted methylation in a wide imprinted region, and possibly identify imprinted genes in the region. *Not*I sites number about 3000 on the whole mouse genome [31,32], and are located at about 1-Mb intervals on average. Similar rare cutter endonucleases used for landmarks have a similar number of recognition sites on the genome as *Not*I. However, because imprinted regions span more than several hundred kbp, we believe our method of screening imprinted loci can detect, with higher probability, loci in imprinted regions.

Two characteristics of imprinted genes were commonly reported beside that of clustered distribution: tandem repeat sequence and a few introns [36,37]. The mouse *U2afbp-rs* gene has both of these characteristics. This gene has a tandem repeat sequence in the 5′ UTR region [34]. The intramolecular homology is shown in Fig. 6.6. On the human homolog *U2AFBPL* on human chromosome 5, this tandem repeat sequence is not found in the upstream region of the gene, and *U2AFBPL* is expressed from both parental alleles in the placenta [38,39]. These results suggest that the tandem repeat sequence has a significant role in imprinting regulation of the *U2afbp-rs/U2AFBPL* gene.

a)

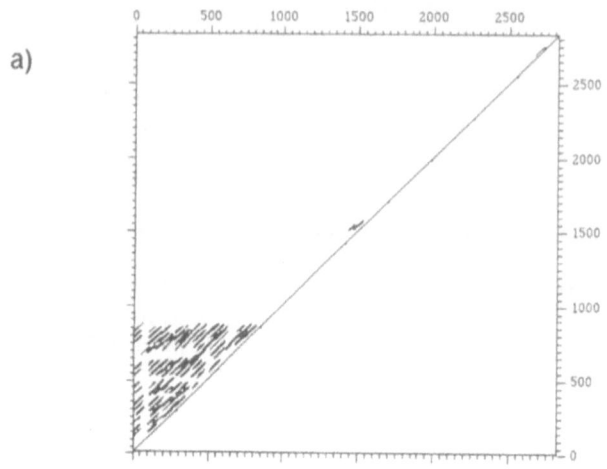

Mouse *U2afbp-rs* cDNA sequence

b)

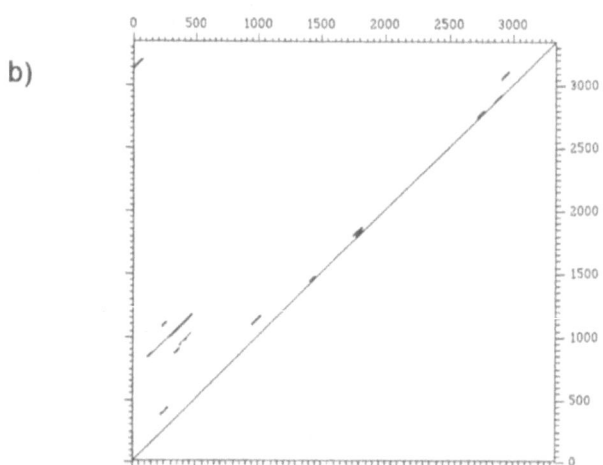

Human U2AFBPL Genomic DNA Sequence

Fig. 6.6a,b. Tandem repeat sequence of *Fok*I family is identified in the promoter region of mouse (**a**) and human (**b**) *U2afbp-rs*, using Harr Plot analysis [38]. (By permission of Academic Press)

Using a strategy of subtraction between cDNA, another screening of imprinted genes was examined [40]. The subtraction was performed in a normally developed embryo and a parthenogenetic embryo. As a result, paternally expressed genes were detected. This subtraction approach targets differentially expressed mRNA.

However, our approach attempted to identify allele-specific methylation on genomic sequences. The mechanism regulating the imprinting phenomena has not yet been elucidated. These different approaches will complementarily contribute toward identifying unknown imprinted genes, thus improving imprinting studies.

References

1. Shibata H, Hirotsune S, Okazaki Y, Komatsubara H, Muramatsu M, Takagi N, Ueda T, Shiroishi T, Moriwaki K, Katsuki M, Chapman VM, Hayashizaki Y (1994) Genetic mapping and systematic screening of mouse endogenously imprinted loci detected with restriction landmark genome scanning method (RLGS). Mammal Genome 5:797–800
2. Shibata H, Yoshino K, Muramatsu M, Plass C, Chapman VM, Hayshizaki Y (1995) The use of RLGS to scan the mouse genome for endogenous loci with imprinted patterns of methylation. Electrophoresis 16:210–217
3. Kawai J, Hirotsune S, Hirose K, Fushiki S, Watanabe S, Hayashizaki Y (1993) Methylation profiles of genomic DNA of mouse developmental brain detected by restriction landmark genomic scanning (RLGS) method. Nucleic Acids Res 21:5604–5608
4. Kawai J, Hirose K, Fushiki S, Hirotsune S, Ozawa N, Hara A, Hayashizaki Y, Watanabe S (1994) Comparison of DNA methylation patterns among mouse cell lines by restriction landmark genomic scanning. Mol Cell Biol 4:7421–7427
5. Lindsay S, Bird A (1987) Use of restriction enzymes to detect potetial gene sequences in mammalian DNA. Nature 327:336–338
6. Crouse H (1960) The controlling element in sex chromosome behaviour in Sciara. Genetics 45:1429–1443
7. Chandra HS, Brown SW (1975) Chromosome imprinting and the mammalian X chromosome. Nature 253:165–168
8. Lyon MF, Rastan S (1984) Parental source of chromosome imprinting and its relevance for X-chromosome inactivation. Differentiation 26:63–67
9. Takagi N, Sasaki M (1975) Preferential inactivation of the paternally derived X chromosome in the extraembryonic membranes of the mouse. Nature 256:640–642
10. Surani MAH, Barton SC (1983) Development of gynecogenetic eggs in the mouse: implications for parthenogenic embryo. Science 222:1034–1036
11. Surani MAH, Barton SC, Norris ML (1984) Development of reconstituted mouse eggs suggests imprinting of the genome during gametogenesis. Nature 308:548–550
12. McGrath J, Solter D (1984) Completion of mouse embryogenesis requires both the maternal and paternal genome. Cell 37:179–183
13. Beechey CV, Cattanach BM (1996) Genetic imprinting map. Mouse Genome 94:96–99
14. Barlow DP, Stöger R, Herrmann BG, Saito K, Schweifer N (1991) The mouse insulin-like growth factor type 2 receptor is imprinted and closely linked to the Tme locus. Nature 349:84–87

15. Nicholls RD, Knoll JHM, Butler MG, Karam S, Lalande M (1989) Genetic imprinting suggested by maternal heterodisomy in non-deletion Prader-Willi syndrome. Nature 342:281–285
16. Searle AG, Edwards JH, Hall JG (1994) Mouse homologues of human hereditary disease. J Med Genet 31:1–19
17. Rainier S, Johnson LA, Dobry CJ, Ping AJ, Grundy PE, Feinberg AP (1993) Relaxation of imprinted genes in human cancer. Nature 362:747–749
18. Ogawa O, Eccle MR, Szeto J, McNoe LA, Yun K, Maw MA, Smith PJ, Reeve AE (1993) Relaxation of insulin-like growth factor II gene imprinting implicated in Wilms' tumor. Nature 362:749–751
19. Sasaki H, Jones PA, Chaillet RJ, Ferguson-Smith AC, Barton SC, Reik W, Surani A (1992) Parental imprinting: potentially active chromatin of the repressed maternal allele of the mouse insulin-like growth factor II(Igf2) gene. Genes Dev 6:1843–1856
20. Stöger R, Kubicka P, Liu C-G, Kafri T, Razin A, Cedar H, Barlow DP (1993) Maternal-specific methylation of the imprinted mouse Igf2r locus identifies the expressed locus as carrying the imprinting signal. Cell 73:61–71
21. Brandeis M, Kafri T, Ariel M, Chaillet JR, McCarrey J, Razin A, Cedar H (1993) The ontogeny of allele-specific methylation associated with imprinted genes in the mouse. EMBO J 12:3669–3677
22. Driscoll DJ, Waters MF, Williams CA, Zori RT, Glenn CC, Avidano KM, Nicholls RD (1992) A DNA methylation imprint, determined by the sex of the parent, distinguishes the Angelman and Prader-Willi syndromes. Genomics 13:917–924
23. Shibata H, Yoshino K, Sunahara S, Gondo Y, Katsuki M, Ueda T, Kamiya M, Muramatsu M, Murakami Y, Kalcheva I, Plass C, Chapman VM, Hayashizaki Y (1996) Inactive allele-specific methylation and chromatin structure of the imprinted gene U2af1-rs1 on mouse chromosome 11. Genomics 35:248–252
24. Kay GF, Barton SC, Surani MA, Rastan S (1994) Imprinting and X chromosome counting mechanisms determine Xist expression in early mouse development. Cell 77:639–650
25. Latham KE, Doherty AS, Scott CD, Schultz RM (1994) Igf2r and Igf2 gene expression in androgenetic, gynecogenetic, and parthenogenetic preimplantation mouse embryos: absence of regulation by genomic imprinting. Genes Dev 8:290–299
26. Szabó PE, Mann JR (1995) Allele-specific expression and total expression levels of imprinted genes during early mouse development: implications for imprinting mechanism. Genes Dev 9:3097–3108
27. Li E, Bestor TH, Jaenisch R (1992) Targeted mutation of the DNA methyltransferase gene results in embryonic lethality. Cell 69:915–926
28. Li E, Beard C, Jaenisch R (1993) Role for DNA methylation in genomic imprinting. Nature 366:362–365
29. Solter D (1988) Differential imprinting and expression of maternal and paternal genomes. Annu Rev Genet 22:127–146
30. Sasaki H, Hamada T, Ueda T, Seki R, Higashinakagawa T, Sakaki Y (1991) Inherited type of allelic methylation variations in a mouse chromosome region where an integrated transgene shows methylation imprinting. Development 111:573–581
31. Antequera F, Bird A (1993) Number of CpG islands and genes in human and mouse. Proc Natl Acad Sci USA 90:11995–11999

32. Imoto H, Hirotsune S, Muramatsu M, Okuda K, Sugimoto O, Chapman VM, Hayashizaki Y (1994) Direct determination of NotI cleavage sites in the genomic DNA of adult mouse kidney and human trophoblast using whole-range restriction landmark genomic scanning. DNA Res 1:239–243

33. Hirotsune S, Shibata H, Okazaki Y, Sugino H, Imoto H, Sasaki N, Hirose K, Okuizumi H, Muramatsu M, Plass C, Chapman VM, Tamatsukuri S, Miyamoto C, Furuichi Y, Hayashizaki Y (1993) Molecular cloning of polymorphic markers on RLGS gel using the spot target cloning method. Biochem Biophys Res Commun 194:1406–1412

34. Hayashizaki Y, Shibata H, Hirotsune S, Sugino H, Okazaki Y, Sasaki N, Hirose K, Imoto H, Okuizumi H, Muramatsu M, Komatsubara H, Shiroishi T, Moriwaki K, Katsuki M, Hatano N, Sasaki H, Ueda T, Mise N, Takagi N, Plass C, Chapman VM (1994) Identification and characterization of an imprinted U2af binding protein related sequence on mouse chromosome 11 detected by efficient genomic screening using Restriction Landmark Genomic Scanning (RLGS-M). Nat Genet 6:33–40

35. Plass C, Shibata H, Kalcheva I, Mullins L, Kotelevtseva N, Mullins J, Sasaki H, Kato R, Hirotsune S, Okazaki Y, Held WA, Hayashizaki Y, Chapman VM (1996) Nat Genet 14:106–109

36. Neumann B, Kubicka P, Barlow DP (1995) Characteristics of imprinted genes. Nat Genet 9:12–13

37. Hurst SD, McVean G, Moore T (1996) Imprinted genes have few and small introns. Nat Genet 12:234–237

38. Pearsall RS, Shibata H, Brozowska A, Yoshino K, Okuda K, Plass C, Chapman VM, deJong PJ, Hayashizaki Y, Held WA (1996) Absence of imprinting in U2AFBPL, a human homologue of the imprinted mouse gene U2afbp-rs. Biochem Biophys Res Commun 222:171–177

39. Kalcheva I, Plass C, Sait S, Eddy R, Shows T, Watkins CD, Camper S, Shibata H, Hayashizaki Y, Ueda T, Takagi N, Chapman VM (1995) Comparative mapping of the imprinted U2afbpL gene on mouse chromosome 11 and human chromosome 5. Cytogenet Cell Genet 68:19–24

40. Kaneko-Ishino T, Kuroiwa Y, Miyoshi N, Kohda T, Suzuki R, Yokoyama M, Viville S, Barton SC, Ishino F, Surani MA (1995) Peg1/Mest imprinted gene on chromosome 6 identified by cDNA sutraction hybridization. Nat Genet 11:25–29

41. Leff SE, Brannan CI, Reed ML, Özçerik T, Francke U, Copeland NG, Jenkins NA (1992) Maternal imprinting of the mouse Snrpn gene and conserved linkage homology with the human Prader-Willi syndrome region. Nat Genet 2:259–264

42. Buiting K, Dittrich B, Gross S, Greger V, Lalande M, Robinson W, Mutirangura A, Ledbetter D, Horsthemke B (1993) Molecular definition of the Prader-Willi syndrome chromosome region and orientation of the SNRPN gene. Hum Mol Genet 2:1991–1994

43. Wevrick R, Kerns JA, Francke U (1994) Identification of a novel paternally expressed gene in the Prader-Willi syndrome region. Hum Mol Genet 3:1877–1882

44. Sutcliffe JS, Nakao M, Christian S, Örstavik KH, Tommerup N, Ledbetter DH, Beaudet AL (1994) Deletions of a differentially methylated CpG island at the SNRPN gene define a putative imprinting control region. Nat Genet 8:52–58

45. Giddings SJ, King CD, Harman KW, Flood JF, Carnaghi LR (1994) Allele specific inactivation of insulin 1 and 2, in the mouse yolk sac, indicates imprinting. Nat Genet 6:310–313
46. Bartolomei MS, Zemel S, Tilghman SM (1991) Parental imprinting of the mouse H19 gene. Nature 351:153–155
47. Zhang Y, Tycko B (1992) Monoallelic expression of the human H19 gene. Nat Genet 1:40–44
48. DeChiara TM, Robertson EJ, Efstratiadis A (1991) Parental imprinting of the mouse insulin-like growth factor II gene. Cell 64:849–859
49. Ohlsson R, Nyström A, Pheifer-Ohlsson S, Töhönen V, Hedborg F, Schofield P, Flam F, Ekström TJ (1993) IGF2 is parentally imprinted during human embryogenesis and in the Beckwith-Wiedemann syndrome. Nat Genet 4:94–97
50. Guillemot F, Caspary T, Tilghman SM, Copeland NG, Gilbert DJ, Jenkins N, Anderson DJ, Joyner AL, Rossant J, Nagy J (1995) Genomic imprinting of Mash2, a mouse gene required for trophoblast development. Nat Genet 9:235–248
51. Hatada I, Mukai T (1995) Genomic imprinting of p57KIP2, a cyclin-dependent kinase inhibitor in mouse. Nat Genet 11:204–206
52. Kalscheuer VM, Mariman EC, Schepens MT, Rehder H, Ropers H-H (1993) The insulin-like growth factor type-2 receptor gene is imprinted in the mouse but not in humans. Nat Genet 5:74–78
53. Villar AJ, Pedersen RA (1994) Parental imprinting of the Mas protooncogene in mouse. Nat Genet 8:373–379
54. Norris DP, Patel D, Kay GF, Penny GD, Brockdorff N, Sheardown S, Rastan S (1994) Evidence that random and imprinted Xist expression is controlled by preemptive methylation. Cell 77:41–51

Chapter 7

Systematic Detection of DNA Alteration in Cancer Tissue

Yasushi Okazaki, Tomoya Ohsumi, Hisato Okuizumi,
William A. Held, and Yoshihide Hayashizaki

Contents

7.1
Overview on Cancer

Tumorigenesis is a multistep process involving both epigenetic and genetic alterations [1,2]. The identification of gain of function mutations in proto-oncogenes and loss of function mutations in tumor suppressor genes has provided a rationale for understanding tumorigenesis. However, the mutation of a single proto-oncogene or tumor suppressor gene is usually not sufficient to cause neoplastic growth. Tumor progression depends on secondary events which arise during cell proliferation. The genetic targets for these secondary events would be expected to depend on the initiating event as well as developmental and tissue-specific factors regulating cell proliferation. Additional steps involving angiogenesis, invasive growth, and metastasis generate more serious life-threatening malignant disease [3]. Although the "cast of characters" involved in these processes is large and growing, our understanding of the process is complicated by the large number of genes involved, developmental and tissue-specific differences in growth regulation, and the stochastic nature of the process.

Alterations in DNA methylation are consistently found associated with tumorigenesis and may play a variety of roles in progression [4,5]. Tumor DNA in general is hypomethylated, which is associated with increased gene expression. Several proto-oncogenes have been shown to be hypomethylated in tumors [4,6]. In contrast, regional hypermethylation may promote tumorigenesis by silencing expression of tumor suppressor genes, genes necessary for differentiation and cessation of growth [4,7,8], or hormone receptors which have a role in growth or properties of the tumor [9,10]. In addition, increased methylation may be mutagenic since 5-methylcytosine can deaminate spontaneously to T, resulting in a C to T transition [4,5]. Results indicating that expression of an exogenous DNA methyltransferase can induce transformation of NIH 3T3 cells support the contention that DNA hypermethylation has an important role in tumor progression [11].

In recent years, a number of new technologies have been developed which can be used for rapidly screening tumors for genetic alterations on a genome-wide basis. Most of these methods depend on nucleic acid hybridization to detect gain or loss of DNA sequences. These methods include comparative genomic hybridization (CGH) [12], representational difference analysis (RDA) [13], PCR analysis of simple sequence length polymorphisms (SSLP) [14], and restriction landmark genomic scanning (RLGS) [15]. Although each method has advantages and disadvantages, the RLGS method is potentially powerful in terms of the number of loci capable of being screened simultaneously, the ability to directly clone genetic regions which are altered during tumorigenesis [16], and the ability to detect and localized regional hypermethylation or hypomethylation changes.

7.2
Introduction

Future progress in the field of cancer research depends greatly on the development of new techniques for the analysis of genomic DNA. A rapid system for analysis of genetic alterations of cancer DNA requires that landmarks and information on chromosomal localization are scanned quickly. Southern blot analysis of restriction fragment length polymorphism (RFLP) and analysis of the simple sequence length polymorphism (SSLP)-PCR system provides a method for mapping of tumor suppressor genes using linkage analysis in families with specific genetic cancer predispositions, as well as by identifying loss of heterozygosity (LOH) in the defined chromosomal regions of sporadic tumors. More than 12 000 markers have been identified [17,18] using SSLP and it has since been employed to analyze DNA in cancer tissue [19]. However, Southern blot analysis and SSLP-PCR are cumbersome and slow in that they require considerable repetition of procedures to insure precise analyses. Since some genetic alterations may occur randomly during tumorigenesis, it is necessary to identify regions of DNA which are frequently lost in a particular type of cancer to identify regions which may contain tumor suppressor genes. Therefore, a system capable of high-speed scanning for a large number of mapped loci must be established.

Using the restriction landmark genomic scanning (RLGS) method, more than 2000 spots/loci throughout the genome can be screened in a single gel. Since an intensive genetic study has identified the chromosomal location of many RLGS spots/loci, it is possible to rapidly determine the chromosomal location of these alterations using NotI, PvuII and PstI. Five hundred and seventy-five C57BL/6 (B6) and Mus spretus (M. spretus) polymorphic loci have been mapped in a single gel. This information can be used to determine if the genetic alterations are contiguous with other loci and thus involve large chromosomal regions.

In this chapter, we describe the application of RLGS to genome-wide analysis of DNA alterations in cancer tissue, employing the liver tumors induced by simian virus 40 T antigen in transgenic mice. To maximize our ability to detect loss at any genetic locus, we have mated the transgeic line (MTD2) in a B6 genetic background with M. spretus to produce interspecific F_1 mice (BSF$_1$) containing one B6 and one M. spretus chromosome. The evolutionary separation between B6 and M. spretus makes it relatively easy to produce RFLPs which can be detected by RLGS.

7.3
Principle and Advantage of the RLGS-Based Screening System in Cancer DNA

Due to the fact that RLGS uses a direct endlabeling system, the spot intensity on RLGS profiles corresponds to the copy number of the cleaved landmark enzyme sites. For diploid RLGS spots, LOH can theoretically be assayed by measuring a decrease in spot intensity. On the basis of this principle, several studies demonstrate that the

RLGS method can identify genetic alterations in the malignant state [20–23]. However, in this type of analysis, it is diffcult, in practice, to detect LOH. This is because LOH would reduce diploid spot intensity by only one half. Moreover, it is not uncommon for tumor tissue to be contaminated with normal tissue, thereby reducing the difference in intensity. However, spots which are polymorphic between B6 and *M. spretus* are haploid, making it easier to detect LOH. The principle of RLGS map-based systematic genome-wide scanning to detect DNA alterations in tumor tissues using the haploid spots is shown in Fig. 7.1. Out of more than 2000 spots on a single RLGS profile, 575 spots (polymorphic between B6 and *M. spretus*) can be analyzed for alterations and localized to specific chromosomal loci. [24,25] (Fig. 7.1).

The major advantages of haploid spots which have been mapped are: (1) the change in spot intensity is much easier to detect with a haploid spot than a diploid; (2) the parental origin of the allele which showed the DNA alteration is distinguished; (3) the polymorphic spots can be subjected to genetic linkage analysis, with the determination of each spot/locus. Information on the chromosomal location of the spots is very useful for screening LOH in tumor DNA. In this study, we used BSF_1 transgenic mice which develop hepatocellular carcinomas for the following reasons

(1) The screening is very efficient due to the fact that about 50% of the spots are polymorphic and therefore haploid;

(2) these polymorphic spots are already mapped in mice, providing very rapid construction of LOH maps, such as the one shown in Fig. 7.3;

(3) in contrast to LOH analysis for humans, this transgenic mouse system offers an unlimited number of tumors of defined stages and types;

(4) this system avoids the complications of genetic heterogeneity;

(5) strain-specific heritable susceptibility effects and parental inheritance effects (imprinting) can also be examined [26,27].

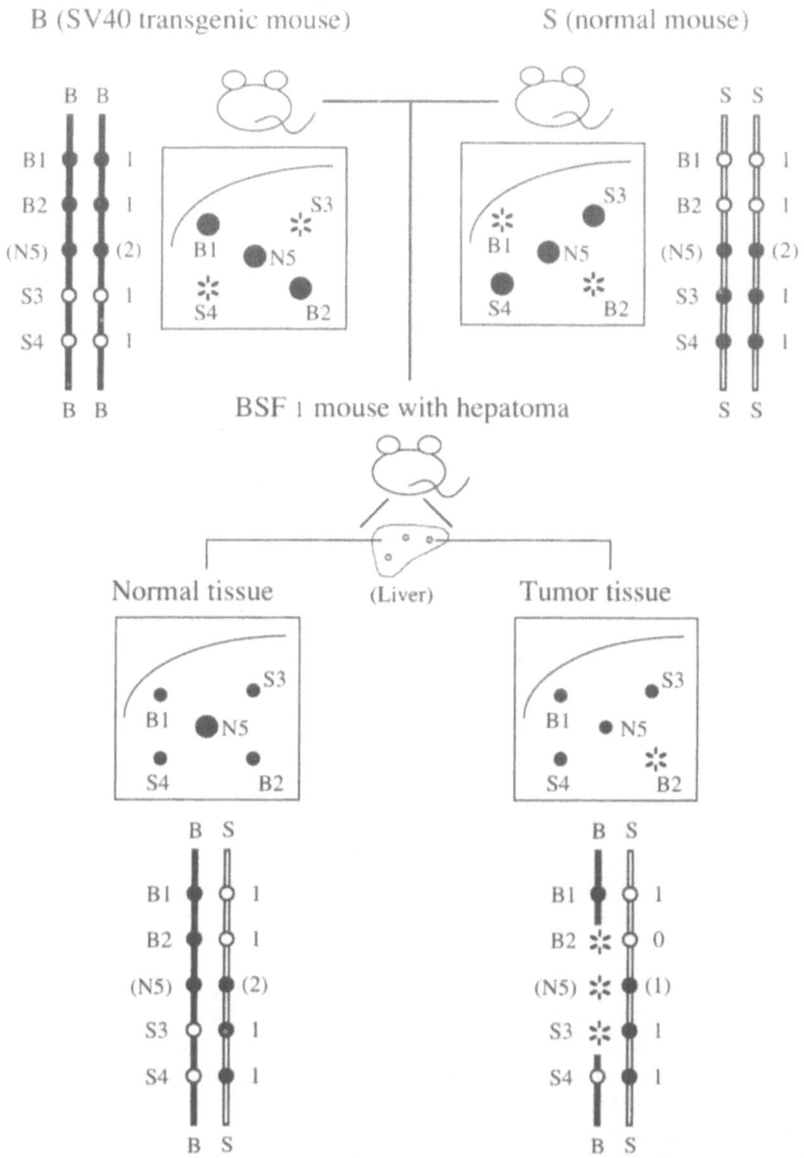

B (SV40 transgenic mouse)   **S (normal mouse)**

Fig. 7.1. Systematic genome-wide scanning to detect DNA alteration in tumor tissues using RLGS genetic map. The map positions of five fictitious loci, B1, B2, S3, S4, and N5, are shown in the progenitor generation. *B1, B2* and *S3, S4* are B- and S-specific spots, respectively. *N5* is a nonpolymorphic spot. The *squares* represent RLGS profile. *Large dots* in the profile represent double intensity, and are homozygous for the presence allele of spots. *Small dots* represent single intensity, and are heterozygous for the loci. The *mark* (*) *of profiles of parents* represents null intensity, and are homozygous for the absence allele. The *mark* (*) *of the profile of tumor tissue* represents chromosomal loss. The *vertical bold and white bars* depict chromosome derived from B (SV40 transgenic B6 mouse) and S (normal *Mus spretus* mouse), respectively. *Black and white circles* on the chromsomes represent presence and absence of allele in the spots, respectively

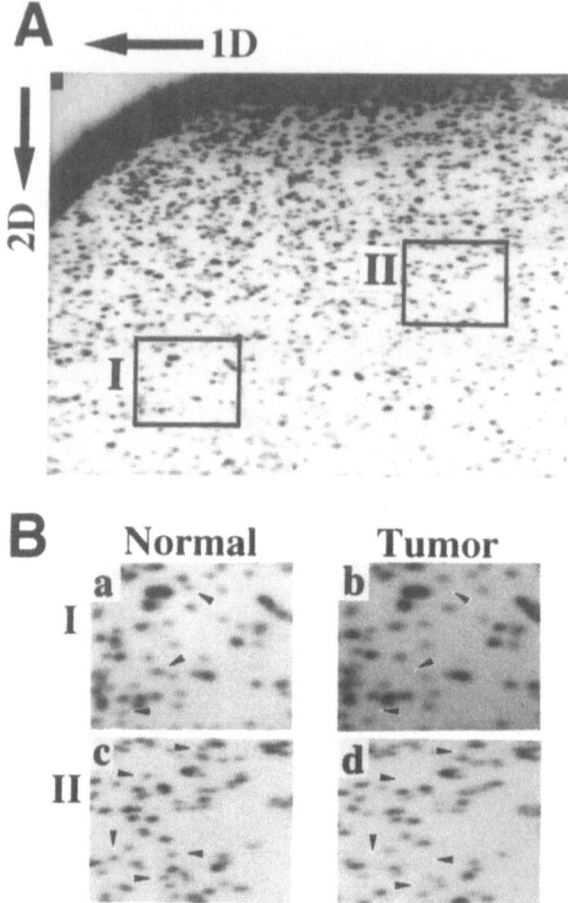

Fig. 7.2A,B. Detection of LOH in liver tumors. **A** RLGS profile of liver tumor B3-2 is shown as a representative. An enzyme combination of *NotI-PvuII-PstI* was used for the analysis. Parts of the RLGS pattern shown in *closed squares* in **A**, designated as *I* and *II*, are magnified in **B**. **B** The *left panels* depict the RLGS profiles in the corresponding area of normal kidney of backcross progeny B3. The *right panels* depict those of tumor tissue. The *arrowheads* indicate RLGS spots which showed disappearance in the tumor profile [16]. (By permission of Academic Press)

7.4
Transgenic Line and Isolation of Liver Tumors

We have previously described [26,27] the MT-D2 transgenic line containing the SV40 early region under the direct control of a mouse major urinary protein (MUP) enhancer/promoter. In this line, multiple tumors develop which have been histopathologically diagnosed as adenomas and hepatocellular carcinomas (HCC). There is complete penetrance of the transgene with all animals developing liver tumors, even when these mice are crossed with various other inbred mouse strains (*M. spretus* included). Male *M. spretus* mice were bred with female MTD2/B6 and the F_1 progeny were sacrificed at 10 to 18 months to obtain liver tumors. Tumor samples were dissected and isolated from adhering normal tissues, along with non-tumor control tissues.

7.5
RLGS-Based System for Detecting LOH
Using Mapped Haploid Spots

Our RLGS map containing 575 haploid spots/loci [340 B6(B)- and 235 *M. spretus*(S)-specific] was generated by performing a single RLGS profile (using the restriction enzymes *Not*I, *Pvu*II, and *Pst*I) on DNA from progeny from a BSS backcross. Since *Not*I is sensitive to DNA methylation of the 5-position of the cytosine residue, RLGS analysis of tumor DNA will detect alterations in the methylation status and the copy number. Virtually all other genomic scanning methods rely on hybridization techniques and therefore can detect only genetic loss. The RLGS method is unique in its ability to detect DNA methylation changes. DNA methylation alterations are commonly associated with tumor progression, where they are thought to have an important role. Thus, the preexisting genetic map of these 575 spots is expected to provide helpful information for the rapid analysis of genomic DNA alterations, such as methylation changes or deletions at landmark sites.

An RLGS profile of transgenic BSF_1 liver tumor with *Not*I, *Pvu*II, and *Pst*I is shown in Fig. 7.2. The arrowheads indicate the spots that decreased in intensity during carcinogenesis. The LOH map of each tumor can be easily drawn after combining the information on the chromosomal location of these 575 spots. Figure 7.3 illustrates the LOH map for a single mouse hepatoma (B3-2) induced by the SV40 transgene. Please note that the two alleles (B6 and *M. spretus*) of each chromosome are separate in this figure. The solid boxes are indicative of regions where more than two spots of decreased intensity in cancer tissue were located contiguously. The region showing contiguous loss appears to reflect LOH due to the regional deletion of chromosome. In addition to the contigu-

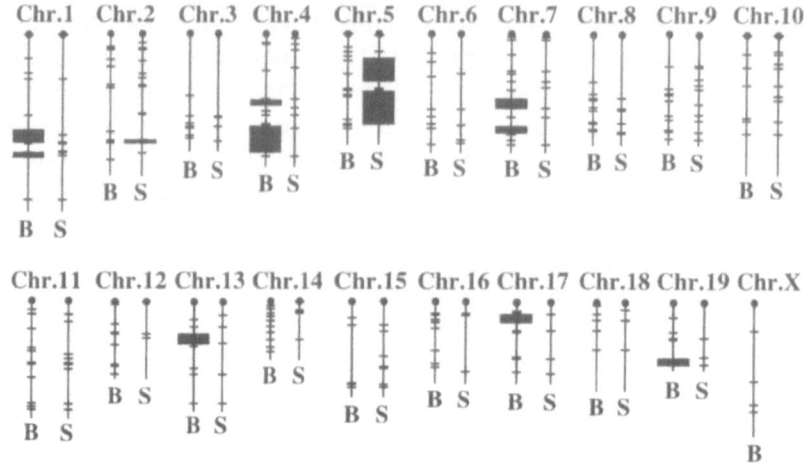

Fig. 7.3. Genome-wide search for LOH in a representative liver tumor. Regions which include at least two sequential spot losses are shown as *solid boxes*. B and S represent the B6 and *M. spretus* allele, respectively. *Horizontal bars* show RLGS markers as previously reported [24,25] [16]. (By permission of Academic Press)

ous loss, solitary spots which decreased in intensity were also found. Most likely these spots are those in which methylation occurred during carcinogenesis.

7.6
Southern Blot Analysis of Cloned RLGS Spots/Loci

In order to determine whether the spot loss was due to a genetic loss (deletion) or to methylation, we cloned 20 spots from solitary and contiguous loss regions. We then tested the genomic DNAs from normal and tumor tissues by Southern blot. We used the same restriction enzyme combination of *Not*I and *Pvu*II as employed in RLGS, because we expected the restriction enzyme fragments to show polymorphisms between B6 and *M. spretus*. *D13Ncvs10* is a typical spot/locus located on the region expected to be the LOH type. The Southern profile using this particular clone is shown in Fig. 7.4A. The *D13Ncvs10* is derived from the B6-specific spot and produces a single unique band in B6 which is the same size as in RLGS, larger than that of *M. spretus* (S). Some tumor tissues (A3-1A and A3-1B) showed extremely weak signals of both B6- and S-specific bands, although the same amount of DNA had been loaded into the slots (note that the β-actin signal has similar intensity in all lanes). This indicates that both alleles were deleted

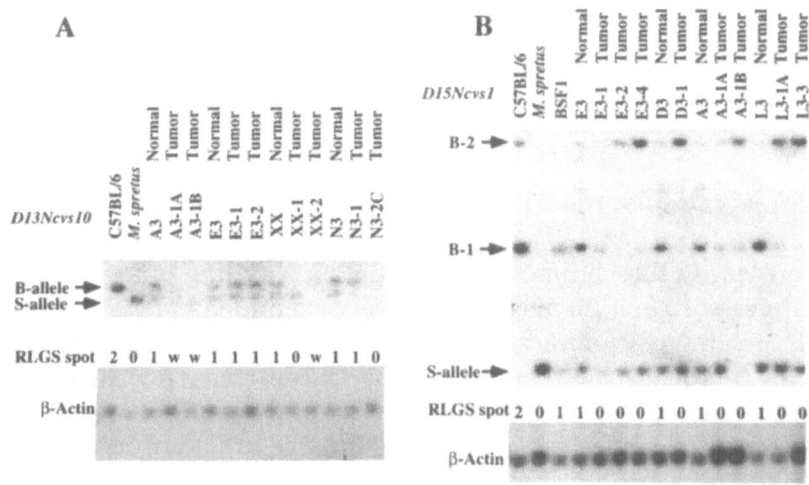

Fig. 7.4A,B. Comparison of Southern blot analysis and RLGS spot analysis. Two representative spots *D13Ncvs10* (**A**) and *D15Ncvs1* (**B**) are shown. These spots were cloned from the punched-out gel. *D13Ncvs10* was picked up from a region showing sequential RLGS spot loss in common with 30 analyzed tumors. *D15Ncvs1* was picked up from a region showing solitary RLGS spot loss. A Southern blot analysis of spot *D13Ncvs10* corresponds to the patterns of RLGS spot analysis shown below. *2, 1, 0,* and *w* represent diploid genome spot intensity, haploid genome spot intensity, loss of spot, and reduced spot intensity (weak), respectively [16]. (By permission of Academic Press)

(i.e., homozygous deletion). The tumors E3-1 and E3-2 showed no DNA alterations. XX-1 produced only a trace of the B6 allele, whereas XX-2 exhibited a deletion of the S allele, although both tumors came from the same mouse liver. Likewise, the tumors N3-1 and N3-2C differed in that N3-1 showed no DNA alteration, while N3-2C showed deletion of both alleles. These results indicate that these tumors are independently derived. *D15Ncvs1* is a typical solitarily B6-specific spot located on the proximal region of Chr 15. In Fig. 7.4B, the B6 allele produces two bands. The upper one (B-2) corresponds to a *Pvu*II fragment which contains the *Not*I landmark site. Thus, this site appears to be partially methylated in normal tissue since the band was not completely digested. The lower band (B-1), which corresponds to the *Not*I-*Pvu*II fragment, will shift to the upper band if the *Not*I site is methylated. At the lower molecular weight is the S-specific band. E3-1 shows the loss of both alleles, since both bands are weak even though the same amounts of DNA were loaded. However, the spots of all remaining tumors decreased in intensity (B-1 band) due to DNA methylation because the B-2 bands were enhanced in these samples. Thus, RLGS analysis of haploid spots can sensitively detect DNA alterations due either to genetic loss or to DNA methylation [28].

7.7
Identification of the LOH Region by RLGS Analysis

To identify the chromosomal region carrying the putative tumor suppressor gene, 30 liver tumors were examined by RLGS for loss of spot intensity indicating LOH. Figure 7.5 shows the map for high frequency LOH in 30 tumors. Chromosomes 5 and 13 are shown as representatives. Because an RLGS spot/locus can only detect B- or S-specifc alleles, we are unable to obtain information on the presence or absence of both alleles at the same locus. In comparison with the SSLP-PCR system, this is somewhat of a drawback.

To circumvent this problem, we assume that in a region showing contiguous loss of more than two B6-specific spots/loci, the B6 allele of S-specific spots within these regions must be lost and vice versa. For example, as shown in the right panel in Fig. 7.5A, tumor B3-2, C3-1A and N3-2C have an S-specific spot loss at *D5Rik133* and *D5Rik122*. In these cases, the S alleles of *D5Ncsv9* or *D5Ncvs10* (these two spots/loci are B6-specific), also located within these regions, were assumed to also be lost. Southern blot analysis using *D5Ncvs9* as a probe confirmed this (data not shown). Based on this assumption, the extensive LOH map was constructed, with Chr. 1(10/30), 5(13/30), 7(11/30), and 13(9/30) exhibiting losses (LOH) of greater than 30% (9/30). These regions are syntenic to human chromosome 1q and 18q (mouse Chr 1), 4p (Chr 5), 11q and 19q (Chr 7), and 5q (Chr 13). There are no known tumor suppressor genes within these regions. These data are relatively consistent with our previous LOH analysis of similar transgenic derived liver tumors which were analyzed by Southern blotting [29].

In addition to mapping the genetic loss regions which involve loss of contiguous RLGS spots/loci, we observed the loss of solitary RLGS spots which are likely to represent alterations in the methylation status as demonstrated for the *D15Ncvs1* locus (see above). There is considerable evidence that tumor cells undergo changes in DNA methylation and that these changes can lead to inactivation of tumor suppressor genes by hypermethylation and possibly to activation of oncogenes by hypomethylation [8,16].

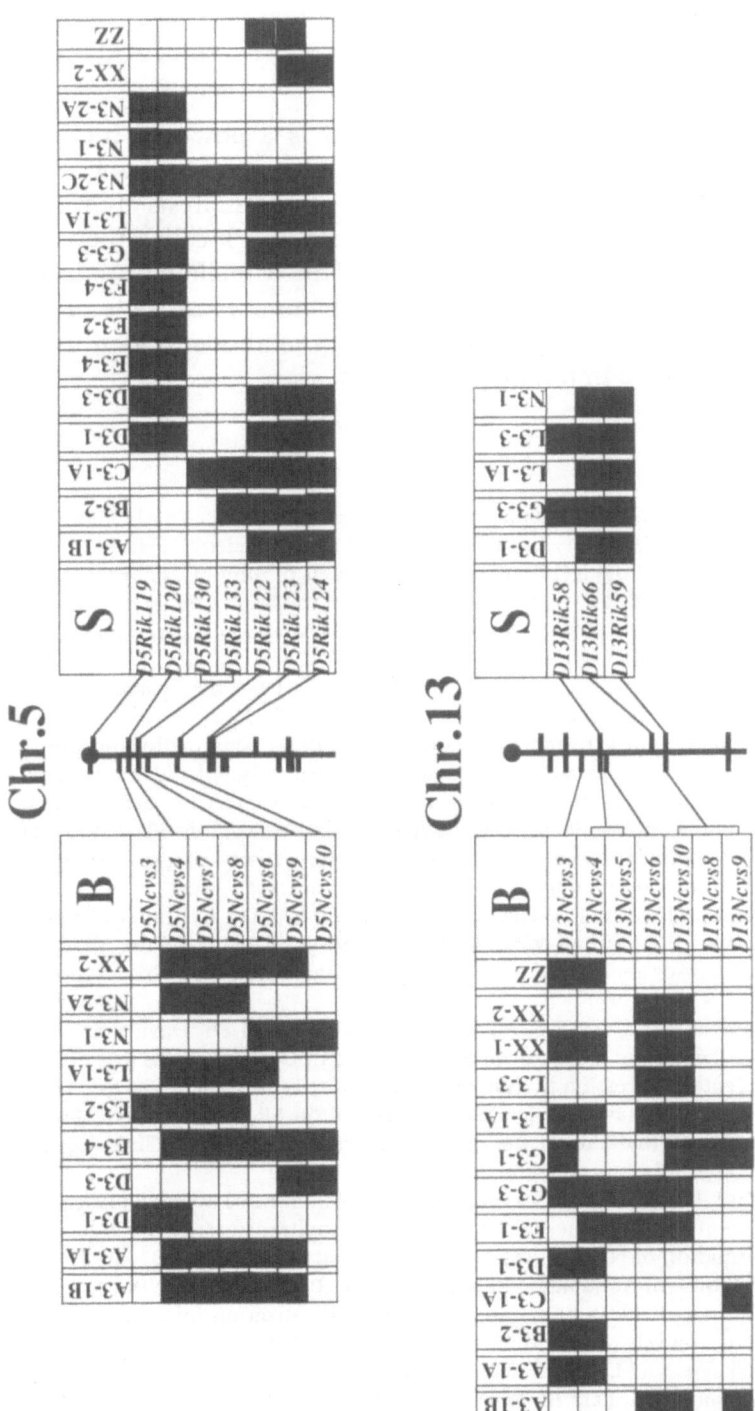

Fig. 7.5A,B. Mapping LOH region using RLGS analysis. LOH regions found in common in 30 tumors are given as a representative. In the names of the markers *D#Ncvs* or *D#Rik*, # indicates the chromosome number. The *solid boxes* indicate the regions where RLGS spots showed a loss. *B* (C57BL/6) and *S* (*Mus spretus*) as in Fig. 7.3

7.8
Conclusion

RLGS map-based screening for LOH is a powerful method for genome-wide scanning for genetic alterations during tumorigenesis. The high-throughput ability of our analysis allowed us to analyze a large number of the tumors and localize the relevant genetic region with some precision. This will allow us to initiate a search for possible candidate tumor suppressor genes in these regions or to utilize positional cloning methods to identify novel genes involved in tumor progression.

References

1. Weinberg RA (1991) Tumor suppressor genes. Science 254:1138–1145
2. Aaronson SA (1991) Growth factors and cancer. Science 254:1146–1153
3. Adams JM, Cory S (1991) Transgenic models of tumor development. Science 254:1161–1167
4. Laird PW, Jaenisch R (1994) DNA methylation and cancer. Hum Mol Genet 3:1487–1495
5. Counts JL, Goodman JI (1995) Alterations in DNA methylation may play a variety of roles in carcinogenesis. Cell 83:13–15
6. Vachtenheim J, Horakova I, Novotna H (1994) Hypomethylation of CCGG sites in the 3′ region of H-ras protooncogene is frequent and is associatedwith H-ras allele loss in non-small cell lung cancer. Cancer Res 54:1145–1148
7. Ohtani-Fujita N, Fujita T, Aoike A, Osifchin NE, Robbins PD, Sakai T (1993) CpG methylation inactivates the promoter activity of the human retinoblastoma tumor-suppressor gene. Oncogene 8:1063–1067
8. Herman JG, Latif F, Weng Y, Lerman MI, Zbar B, Liu S, Samid D, Duan D-SR, Gnarra JR, Linehan WM, Baylin SB (1994) Silencing of the VHL tumor-suppressor gene by DNA methylation in renal carcinoma. Proc Natl Acad Sci USA 91:9700–9704
9. Ottaviano YL, Issa J-P, Parl FF, Smith HS, Baylin SB, Davidson NE (1994) Methylation of the estrogen receptor gene CpG island marks loss of estrogen receptor expression in human breast cancer cells. Cancer Res 54:2552–2555
10. Issa J-PJ, Ottaviano YL, Celano P, Hamilton SR, Davidson NE, Baylin, SB (1994) Methylation of the oestrogen receptor CpG island links ageing and neoplasia in human colon. Nat Genet 7:536–540
11. Wu J, Issa J-P, Herman J, Bassett DE Jr, Nelkin BD, Baylin SB (1993) Expression of an exogenous eukaryotic DNA methyltransferase gene induces transformation of NIH 3T3 cells. Proc Natl Acad Sci USA 90:8891–8895
12. Kallioniemi A, Kallioniemi O-P, Sudar D, Rutovitz D, Gray JW, Waldman F, Pinkel D (1992) Comparative genomic hybridization for molecular cytogenetic analysis of solid tumors. Science 258:818–821
13. Lisitsyn NA, Lisitsina NM, Dalbagni G, Barker P, Sanchez CA, Gnarra J, Linehan WM, Reid BJ, Wigler MH (1995) Comparative genomic analysis of

tumors: detection of DNA losses and amplification. Proc Natl Acad Sci USA 92:151–155

14. Dietrich W, Katz H, Lincoln SE, Shin H-S, Friedman J, Dracopoli NC, Lander ES (1992) A genetic map of the mouse suitable for typing intraspecific crosses. Genetics 131:423–447

15. Hatada I, Hayashizaki Y, Hirotsune S, Komatusbara S, Mukai T (1991) A genomic scanning method for higher organisms using restriction sites as landmark. Proc Natl Acad Sci USA 88:9523–9527

16. Ohsumi T, Okazaki Y, Okuizumi H, Shibata K, Hanami T, Mizuno Y, Takahara T, Sasaki N, Ueda M, Muramatsu M, Kerns KA, Chapman VM, Held WA, Hayashizaki Y (1995) Loss of hetrozygosity in chromosome 1, 5, 7 and 13 in mouse hepatoma detected by systematic genome-wide scanning using RLGS genetic map. Biochem Biophys Res Commun 212:632–639

17. Dietrich WF, Miller J, Steen R, Merchant MA, Damron-Boles D, Husain Z, Dredge R, Daly MJ, Ingalls KA, O'Connor TJ, Evans CA, DeAngelis MM, Levinson DM, Kruglyak L, Goodman N, Copeland NG, Jenkins NA, Hawkins TL, Stein L, Page DC, Lander ES (1996) A comprehensive genetic map of the mouse genome. Nature 380:149–152

18. Dib C, Faure S, Fizames C, Samson D, Drouot N, Vignal A, Millasseau P, Marc S, Hazan J, Seboun E, Lathrop M, Gyapay G, Morissette J, Weissenbach J (1996) A comprehensive genetic map of the human genome based on 5,264 microsatellites. Nature 380:152–154

19. Dietrich WF, Radany EH, Smith JS, Bishop JM, Hanahan D, Lander ES (1994) Genome-wide search for loss of heterozygosity in transgenic mouse tumors reveals candidate tumor suppressor genes on chromosomes 9 and 16. Proc Natl Acad Sci USA 90:9451–9455

20. Hayashizaki Y, Hirotsune S, Okazaki Y, Hatada I, Shibata H, Kawai J, Hirose K, Watanabe S, Fushiki S, Wada S, Sugimoto T, Kobayakawa K, Kawara T, Katsuki M, Sibuya T, Mukai T (1993) Restriction landmark genomic scanning method and its various applications. Electrophoresis 14:251–258

21. Hirotsune S, Hatada I, Komatsubara H, Nagai H, Kanji K, Kobayakawa K, Kawara T, Nakagawara A, Fujii K, Mukai T, Hayashizaki Y (1992) New approach for detection of amplification in cancer DNA using Restriction Landmark Genomic Scanning. Cancer Res 52:3642–3647

22. Nagai H, Ponglikitmongkol M, Mita E, Ohmachi Y, Yoshikawa H, Saeki R, Yumoto Y, Nakanishi T, Matsubara K (1994) Aberration of genomic DNA in association with human hepatocellular carcinomas detected by 2-dimensional gel analysis. Cancer Res 54:1545–1550

23. Miwa W, Yashima K, Sekine T, Sekiya T (1995) Demethylation of a repetitive DNA sequence in human cancers. Electrophoresis 16:227–232

24. Hayashizaki Y, Hirotsune S, Okazaki Y, Shibata H, Akasako A, Muramatsu M, Kawai J, Hirasawa T, Watanabe S, Shiroishi T, Moriwaki K, Taylor BA, Matsuda Y, Elliott RW, Manly KF, Chapman VM (1994) A genetic linkage map of the mouse using Restriction Landmark Genomic Scanning (RLGS). Genetics 138:1207–1238

25. Okuizumi H, Okazaki Y, Ohsumi T, Hanami T, Mizuno Y, Muramatsu M, Hayashizaki Y, Plass C, Chapman VM (1995) A single gel analysis of 575 dominant and codominant restriction landmark genomic scanning loci in mice interspecific backcross progeny. Electrophoresis 16:253–260

26. Held WA, Mullins JJ, Kuhn NJ, Gallagher JF, Gu GD, Gross KW (1989) T antigen expression and tumorigenesis in transgenic mice containing a mouse major urinary protein/SV40 T antigen hybrid gene. EMBO J 8:183–191
27. Schirmacher P, Held WA, Yang D, Biempica L, Rogler CE (1991) Selective amplification of periportal transitional cells precedes formation of hepatocellular carcinoma in SV40 large tag transgenic mice. Am J Pathol 139:231–241
28. Akama TO, Okazaki Y, Ito M, Okuizumi H, Konno H, Muramatsu M, Plass C, Held WA, Hayashizaki Y (1997) RLGS-M based genome-wide scanning of mouse liver tumors for alterations in DNA methylation status. Cancer Res, in press
29. Held WA, Pazik J, O'Brien JG, Kerns K, Gobey M, Meis R, Kenny L, Rustum Y (1994) Genetic analysis of liver tumorigenesis in SV40 T antigen transgenic mice implies a role for imprinted genes. Cancer Res 54:6489–6495

Chapter 8

RLCS, Restriction Landmark cDNA Scanning

TAKESHI YAOI, HARUKAZU SUZUKI, JUN KAWAI, and
SACHIHIKO WATANABE

Contents

8.1
Introduction

It is important to identify and isolate differentially expressed genes whose expression underlies many biological processes such as development, differentiation, and cellular response to various stimuli, and also pathological changes that arise in diseases. Classically, two hybridization methods, differential and subtractive hybridization, have been used to analyze and isolate such genes [1–3]. However, differential hybridization is effective only for mRNAs expressed abundantly in one of the two types of cells. Subtractive hybridization is rather empirical and poor in reproducibility. Furthermore, both methods are time-consuming. Recently, Liang and Pardee have developed a novel method called differential display using an arbitrarily primed reverse transcription-coupled PCR and sequencing gels [4]. This has proved to be a powerful technique for detecting changes in eukaryotic gene expression; it is simple, sensitive, and time-saving. However, PCR-mediated amplification with arbitrary primers requires delicate control, depending on the primers used [5,6], and cDNA probes thus obtained often give false positive signals on Northern blot analysis [7,8]. Furthermore, the differential display seems less sensitive in detecting rare mRNA populations, since it has a strong bias toward higher abundance mRNAs [9].

The RLGS (restriction landmark genomic scanning) is a novel method for genomic DNA scanning in which many loci can be visualized and quantified as two-dimensional gel spots [10,11]. It has proved to be very powerful in various applications such as genetic mapping [12,13], systematic detection of methylatable loci including imprinted genes [11,14–19], and systematic detection of aberrated DNA in cancer cells [11,20–22]. Furthermore, simple methods have also been established for cloning the target DNA fragments from RLGS gel spots [23,24]. Here, we introduce an alternative cDNA display system, designated RLCS (restriction landmark cDNA scanning), by which the RLGS can be applied to display many cDNA species as two-dimensional gel spots [25].

8.2
Principle

Essentially, the principle of RLGS is adopted for RLCS. The cDNA fragments with cleaved restriction sites which are labeled as landmarks are separated by high-resolution two-dimensional gel electrophoresis followed by autoradiography. However, in RLCS, it is very important to prepare cDNA species of a length uniform with each mRNA species. For this purpose we designed a particular olig(dT) anchor primer for cDNA synthesis, which can anchor to mRNA at the upstream end of the poly(A)$^+$ tail, allowing the cDNA products to be recovered easily. The outline of RLCS is shown in Fig. 8.1A and B. The primer has a biotin residue at the 5′-end, multiple restriction enzyme sites including NotI, a 15-mer dT stretch, and an additional 2nt MN at the 3′-end (M = A, G, or C; N = A, G, C, or T) as indicated in Fig. 8.2. The primer whose N is A was used in the experiment described below, but in principle different cDNA species will be displayed by selecting a different base as N. cDNA is synthesized with the cDNA primer and blocked by ddNTPs to prevent non-specific labeling in the following step (Fig. 8.1A: steps a and b). The cDNA synthesized is then digested with restriction enzyme A as described in RLGS. The enzyme which creates proturding cohesive 5'-termini is preferable as enzyme A (Fig. 8.1A, step c), because its restriction sites are radiolabeled by Sequenase (Fig. 8.1A step d) and also advantageous for spot cloning (Fig. 8.4). The labeled cDNA fragments downstream of the enzyme sites should be uniform in length for individual mRNA species. Thus, the fragments are recovered using streptavidin-conjugated magnetic beads (Fig. 8.1A, step e), and released from the beads by *Not*I treatment (Fig. 8.1A, step f).

The cDNA samples obtained above are applied to high-resolution two-dimensional gel electrophoresis as previously described in RLGS (Fig. 8.1B). After first-dimension agarose gel electrophoresis, the cDNA fragments in the gel are completely digested in situ with restriction enzyme B, and subjected to second-dimension acrylamide gel electrophoresis. Then the gel is dried, followed by autoradiography. Consequently, numerous individual cDNA species are displayed as gel spots according to the positions of the enzyme A site and its nearest downstream enzyme B site.

8.3
Protocol for RLCS

8.3.1
Preparation of Total RNA

Total RNA from tissues or cells is prepared by the AGPC method [26]. For one successful cDNA synthesis it is recommended to isolate at least 500 µg of total RNA in one experiment. This yield is obtained from approximately 10^8 cells when cells such as PC12, for example, are used as starting materials.

8.3.2
Poly(A) RNA Purification, cDNA Synthesis, Labeling, and 2-D Gel Electrophoresis

High quality poly(A) RNA is essential to obtain satisfactory RLCS patterns. After publication of our initial report on RLCS [25], the poly(A) RNA preparation was slightly modified to reduce the background signal as described below.

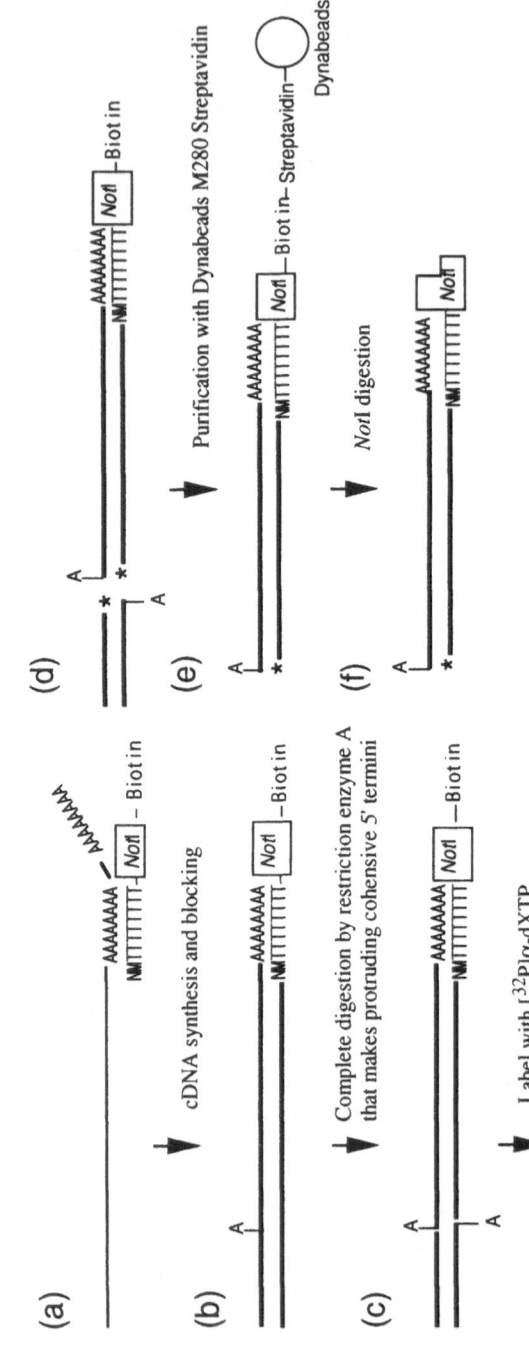

Fig. 8.1. Schematic representation of the principle of RLCS. **A** Preparation of RLCS sample. *Thin horizontal line* and *thick double horizontal lines* indicate poly(A)⁺ RNA and double-stranded cDNA, respectively. **B** Separation of cDNA species by two-dimensional gel electrophoresis. cDNA fragments in an RLCS sample (*CS1*, *CS2*, and *CS3*) are separated by two-dimensional gel electrophoresis and detected as spots CS1, CS2, and CS3, respectively. From H. Suzuki et al. [25] by permission of Oxford University Press

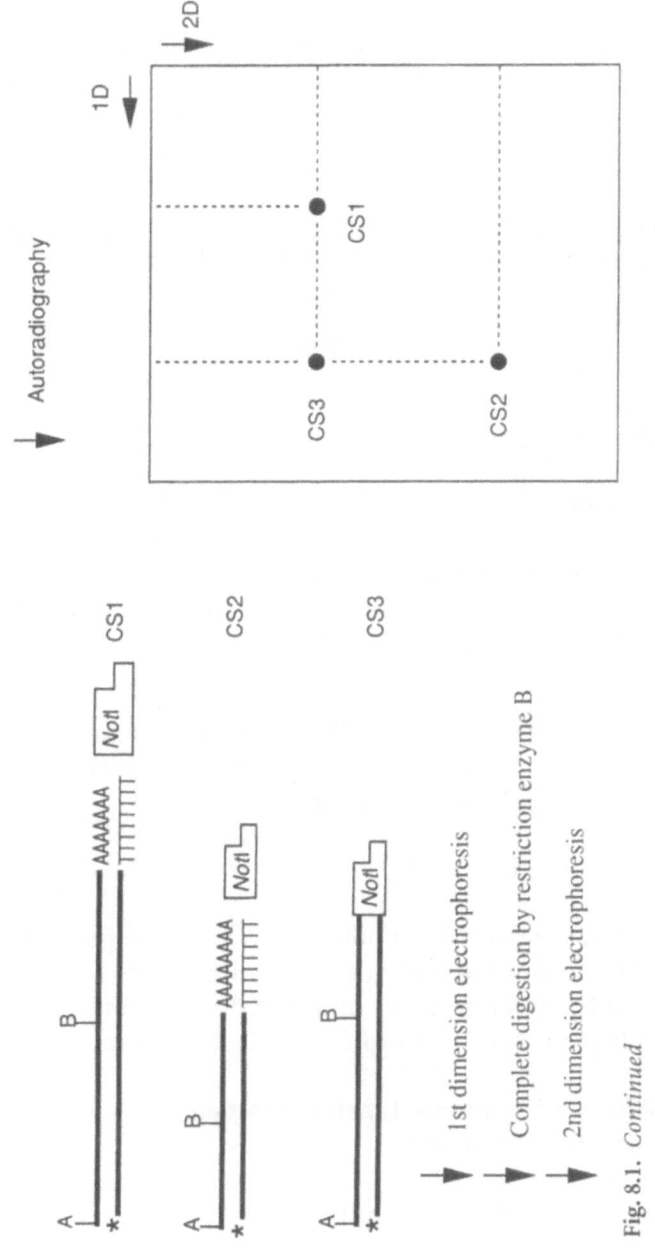

Fig. 8.1. *Continued*

Anchor primer

3'-AM(T)₁₅CC<u>CGCCGGCG</u>AGCGCT<u>AGATCT</u>TGATCAG-Biotin

<small>M=A or G or C</small>   *Not*I   *Nru*I   *Xba*I   *Spe*I

3'-GM(T)₁₅CC<u>CGCCGGCG</u>AGCGCTAGATCTTGATCAG-Biotin
3'-CM(T)₁₅CC<u>CGCCGGCG</u>AGCGCTAGATCTTGATCAG-Biotin
3'-TM(T)₁₅CC<u>CGCCGGCG</u>AGCGCTAGATCTTGATCAG-Biotin

Fig. 8.2. Biotinylated anchor primers designed for RLCS

Purification of Poly(A) RNA

We usually purify mRNAs from 500 µg of total RNA in one experiment. To minimize ribosomal RNA contamination in poly(A) RNA preparation we routinely use the magnetic beads conjugated with dT-oligonucleotides such as Dynabeads(dT)25 (DYNAL, Inc.) or BioMag Oligo (dT)20 (PerSeptive Diagnostics, Inc.) according to the manufacturer's instruction manual on purification.

Note: Poly(A) RNAs are usually recovered at the following percentages in total RNAs in using such magnetic beads: 1.5–2% for brain tissue or 1–2% for PC12 cells.

cDNA Synthesis, Labeling Reaction, and 2-D Gel Electrophoresis

For cDNA synthesis, 5 µg of poly(A) RNA is used in each reaction. The cDNA synthesized is divided into four fractions, each of which is digested by different types of restriction enzymes, labeled, and then applied to two-dimensional gel electrophoresis.

Material for cDNA Synthesis, Labeling Reaction and 2-D Gel Electrophoresis

Kit – SuperScript Choice System (BRL)

Enzymes – SuperScriptII RNase H⁻ reverse transcriptase(200 U/µl) (BRL)

  – *E. coli* DNA ligase(10 U/µl) (BRL)

  – DNA polymerase I(5–10 U/µl) (BRL)

  – ribonuclease H(1–4 U/µl) (BRL)

- RNase ONE Ribonuclease(5–10 U/µl) (Promega)

- Sequenase (Ver.2.0) (Amersham)

- restriction enzyme for the cDNA digestion (5′-protruding 6-base cutter; *Bam*H I, *Bgl*II, *Nco*I, *Eco*RI, *Hin*dIII, *Bsp*HI, *Bln*I, and so on)

- *Not*I

- restriction enzyme for the 2nd cDNA digestion (for in situ gel digestion) (5-base cutter; *Hin*fI has proved convenient)

- 5× first-strand buffer (BRL) [included in the kit (SuperScript Choice System) and SuperScriptII]   **Buffers**

- 5× second-strand buffer (BRL) [included in the kit (SuperScript Choice System)]

- 10× RNase ONE buffer (Promega) [appended to RNase One]

- 10× T buffer [330 mM Tris-acetate (pH 7.9)/100 mM Mg-acetate/5 mM DTT/660 mM K-acetate]

- 10× Klenow buffer

- STE-BSA buffer [10 mM Tris-Cl (pH 7.5)/1 mM EDTA (pH 8.0)/0.5 M NaCl/0.01% BSA]

- 10× 1-D electrophoresis buffer

Trizma Base	121.0 g
sodium acetate 3H$_2$O	54.5 g
NaCl	21.0 g
EDTA-Na$_2$-H$_2$O	11.7 g

fill up to 1 l with DW to ca. pH 8.1 with acetic acid (ca. 40 ml)

Note: Stock the buffer at 4 °C, keeping it not longer than 3 months. Deterioration of the buffer is one of the causes of smearing in the spot pattern.

- restriction enzyme reaction buffer for the 2nd cDNA digestion (refer to manufactures' instruction except for addition of final 0.01% BSA to reaction mixture and omission of DTT)

- Chromaspin TE-100 column (CLONTECH)   **Spun column**

- 0.1 M DTT (BRL) [included in the kit (SuperScript Choice **Reagents** System) and SuperScript II]

- 10 mM dNTP mix (BRL) (included in the kit)

- 1 mM ddGTP, 1 mM ddATP, 1 mM ddTTP, 1 mM ddCTP

- [α-^{32}P] dXTP(6000 Ci/mmol) [X = G or A or C] (Amersham, DuPONT)

- DEPC-treated water

- Dynabeads M-280 Streptavidin (Dynal)

Preparation of Dynabeads M-280 Streptavidin:

1. Add the appropriate amounts of resuspended Dynabeads M-280 streptavidin into a 1.5 ml tube.

2. Place the tube in Dynal Magnetic particle concentrator (Dynal MPC) for 1–2 min.

3. Carefully remove the supernate by aspiration with a pipette while the tube remains in Dynal MPC.

4. Remove the tube from Dynal MPC. Add to the tube STE-BSA buffer with the same volume as in step 1. Resuspend gently.

5. Repeat steps 3 and 4.

6. Repeat steps 3 and 4. The working suspension of Dynabeads M-280 is obtained. Resuspend it immediately before use.

- TE-saturated phenol/chloroform/isoamylalcohol(50:49:1)

- 70% ethanol

- 1% acrylmonomer [preparation: add 100 µl 10% APS and 10 µl of TEMED to 10 ml 5% acrylamide (monomer) in distilled water. Polymerize at room temperature for 20–30 min. Dilute to 1% with distilled water.]

- Sea-Kem GTG agarose (FMC Inc.)

- 6× BPB dye (0.25% bromophenol blue/40% w/v sucrose in water. Store at 4 °C)

Note: Purchase of cDNA synthesis reagents. To provide reagents, it is more convenient to purchase the kit. When any constituent like enzyme or dNTP mix in the kit is lacking, it is commercially available (for example, from BRL) except for 5× second-strand buffer.

– 5'-biotinylated cDNA synthesis primer **Sequence**

Sequence: 5' – Biotin-GAC-TAG-TTC-TAG-ATC-GCG-AGC-GGC-CGC-CCT-TTT-TTT-TTT-TTT-TTMN (M = G,A,C; N = G or A or T or C) – 3'

Protocol for cDNA Synthesis, Labeling, and 2-D Gel Electrophoresis

Our routine protocol is shown schematically as an experimental flow chart. Here, double-stranded cDNA synthesis reaction is based on the instruction manual in the BRL kit (SuperScript Choice System). If necessary, refer to the manual for further information. For the use of Chromaspin TE-100 spin column, see also the manufacturer's manual. Take great care in labeling, since it is very important to exclude possible background signals. The detailed protocol for 2-D gel electrophoresis is described in Section 3.2.2.

cDNA Synthesis

Experimental protocol

1. Add 5 µl of poly(A)$^+$ RNA (1 µg/µl) and 2 µl of biotinylated primer (0.5 µg/µl) to an RNase-free 1.5 ml tube. Mix well with pipette.

2. Incubate the mixture at 70 °C for 10 min, and then chill quickly on ice. Collect the mixture by brief centrifugation.

3. Add the follwing: 1 µl of DEPC-treated water, 4 µl of 5× 1st-strand buffer, 2 µl of 0.1 mM DTT, and 1 µl of 10 mM dNTPs. Mix gently with pipette.

4. Incubate at 37 °C for 2 min.

5. Add 5 µl of SuperScript II. Mix gently with pipette. Collect the mixture by brief centrifugation.

6. Incubate at 37 °C for 1 h.

7. Place the tube on ice.

8. Add the follwing reagents in the order: 93 µl of DEPC-treated water, 30 µl of 5× 2nd-strand buffer, 3 µl of 10 mM dNTPs, 1 µl of *E. coli* DNA ligase (10 U/µl), 4 µl of *E. coli* DNA polymerase I (10 U/µl), and 1 µl of *E. coli* RNase H (2 U/µl). Mix gently. Collect the mixture by brief centrifugation.

9. Incubate at 16 °C for 2 h.

10. Add 10 μl of 0.5 M EDTA (pH 8.0) to terminate the reaction. Vortex thoroughly and centrifuge briefly.

11. Add 150 μl of TE-saturated phenol/chloroform/isoamylalcohol, vortex thoroughly, and centrifuge at room temperature for 5 min at $14\,000 \times g$.

12. Recover 145 μl of upper aqueous layer to a fresh tube.

13. Add 1.5 μl of 1% linearized acrylamide, 14.5 μl of 3 M NaOAc, and 145 μl of isopropanol. Vortex thoroughly and centrifuge at 4 °C for 5 min at $14\,000 \times g$.

14. Discard the supernatant. Add 150 μl of cold ethanol to precipitate. Centrifuge at 4 °C for 1 min at $14\,000 \times g$.

15. Discard the supernatant. Dry the pellet at 37 °C for 5 min to evaporate residual ethanol.

RNase Treatment

1. Dissolve the cDNA pellet with 50 μl of 1× RNase ONE buffer.

2. Add 1 μl of RNase ONE (10 U/μl). Mix thoroughly and centrifuge briefly.

3. Incubate at 37 °C for 20 min.

4. Add 45 μl of TE-saturated phenol/chloroform/isoamylalcohol, vortex thoroughly, and centrifuge at room temperature for 5 min at $14\,000 \times g$.

5. Recover 50 μl of upper aqueous layer.

6. Apply it to the spun column, Chromaspin TE-100, to remove the digested RNA (according to manufacturer's instruction manual).

7. To estimate the size distribution of synthesized cDNA, transfer 1 μl of the eluted cDNA (= x μl) to a fresh tube which contains 1 μl of 6× BPB dye and 4 μl of DW, and apply to 1% agarose gel electrophoresis. If the size distribution is acceptable, proceed to next step.

8. To the remaining cDNA solution [(x − 1)μl], add x/100 μl of 1% linearized acrylamidgb e, x/10 μl of 3 M NaOAc, and x μl of

isopropanol. Vortex thoroughly and centrifuge at 4 °C for 5 min at $14\,000 \times g$.

9. Discard the supernatant and add 150 µl of cold ethanol. Centrifuge at 4 °C for 1 min at $14\,000 \times g$.

10. Discard the supernatant. Dry the pellet at 37 °C for 5 min to evaporate residual ethanol.

11. Redissolve the pellet with 20 µl of 10-fold diluted TE buffer.

Note: After ds-cDNA synthesis, the cDNA is treated with RNase and digested RNA is removed. This step is essential to reduce the background noise on RLCS profiles.

Blocking

1. Add the follwing reagents to the cDNA solution (20 µl) and mix gently: 5 µl of 10× Klenow buffer, 5 µl each of ddNTPs, 5 µl of 0.1 M DTT, 1 µl of Sequenase (Ver. 2.0). Centrifuge the mixture briefly.

2. Incubate at 37 °C for 30 min.

3. Add 45 µl of TE-saturated phenol/chloroform/isoamylalcohol, vortex thoroughly, and centrifuge at room temperature for 5 min at $14\,000 \times g$.

4. Recover 50 µl of upper aqueous layer.

5. Apply to Chromaspin TE-100 to remove unincorporated ddNTPs.

6. Divide the eluted cDNA into quarters.

Note: Each (y µl) of these aliquots can be treated with a different 'restriction enzyme A' as shown in the next step.

Digestion

1. Add the following reagents to a quarter volume (y µl) of the above cDNA: y/7 µl of 10× T buffer, y/7 µl of 40 mM spermidine, ≤y/7 µl of 'restriction enzyme A'. Fill a total of 10y/7 µl with DW, mix thoroughly, and spin down briefly.

9y/7 µl

y/7 µl

Use y/7 µl of the mixture as the control for cut check.

Add 2 µl of λDNA solution (0.5 µg/µl; in 1× T buffer/ 4 mM spermidine).

Mix gently with pipette. (We call this the 'control mixture').

Mix the following mixture 'λ-enzyme A' in another fresh tube (use this mixture as a size marker in the following gel electrophoresis).

2 µl of λDNA (0.5 µg/µl), 0.5 µl of 10× T buffer, 0.5 µl of 40 mM spermidine, 0.5 µl of 'restriction enzyme A', and 1.5 µl of DW.

2. Incubate at 37 °C for 1.5 h.

Incubate both mixtures at 37 °C for 1.5 h. 1% agarose gel electrophoresis (compare the digested patterns of λDNA to each other. If 'control mixture' is completely digested, go to step 3).

3. Fill the digested cDNA (9y/7 µl) to a total of 50 µl with DW.

4. Add 45 µl of TE-saturated phenol/chloroform/isoamylalcohol to it.

5. Vortex thoroughly and centrifuge at room temperature for 5 min at $14\,000 \times g$.

6. Recover 50 µl of upper aqueous layer to a fresh tube.

7. Add 0.5 µl of 1% linearized acrylamide, 5 µl of 3 M NaOAc, and 50 µl of isopropanol.

8. Vortex thoroughly and centrifuge at 4 °C for 5 min at $14\,000 \times g$.

9. Discard the supernatant and add 150 µl of cold ethanol. Centrifuge at 4 °C for 1 min at 14 000 × g.

10. Discard the supernatant. Dry the pellet at 37 °C for 5 min to evaporate residual ethanol.

11. Dissolve the pellet with 5.5 µl of DW.

Labeling and Purification of the cDNA

1. Prepare 10-fold diluted Sequenase (ver. 2.0) which contains 1 µl of 10× Klenow buffer, 1 µl of 0.1 M DTT, 7 µl of DW and 1 µl of Sequenase (ver. 2.0). Keep on ice.

2. Add the following reagents to the digested cDNA: 1.5 µl of 10× Klenow buffer, 1.5 µl of 0.1 M DTT, 1.5 µl of 1 mM ddCTP, 1.5 µl of 1 mM ddTTP, 2.5 µl of $[\alpha^{32}P]$-dGTP(6000 Ci/mmol), and 1 µl of 10-fold diluted Sequenase (ver. 2.0) (step 1).

3. Mix gently with pipette.

4. Keep at room temperature for 5 min.

Note: Reaction time is critical to avoid nonspecific labeling.

5. While keeping on ice, add immediately 1.2 µl of 0.1 M EDTA (pH 8.0).

6. Add the following reagents in the order: 1.2 µl of Tris-Cl (pH 7.5), 70.6 µl of DW, 12 µl of 5 M NaCl, and 20 µl of Dynabeads M-280.

7. Mix thoroughly with pipette.

8. Incubate at 37 °C for 30 min with gentle rotation or occasional mixing.

9. Separate the Dynabeads M-280 by putting the tube in a Dynal Magnetic Particle Concentrator (Dynal MPC) for 1 to 2 min. Discard the supernatant while keeping the tube in the Dynal MPC.

10. Resuspend the pellet in 150 µl of STE-BSA buffer.

11. Put the tube in the Dynal MPC for 1 to 2 min. Discard the supernatant while keeping the tube in the Dynal MPC.

12. Resuspend the pellet in 150 µl of STE-BSA buffer.

13. Put the tube in the Dynal MPC for 1 to 2 min. Discard the supernatant with keeping the tube in the Dynal MPC.

14. Resuspend the pellet in 150 µl of 1× high buffer/0.01% BSA.

15. Put the tube in the Dynal MPC for 1 to 2 min.

16. Discard the supernatant while keeping the tube in the Dynal MPC.

17. Resuspend the pellet in 200 µl of 1× high buffer/Not I (75 U/ 200 µl).

18. Incubate at 37 °C for 30 min with gentle rotation or occasional mixing.

19. Put the tube in the Dynal MPC for 1 to 2 min.

20. Recover the supernatant while keeping the tube in the Dynal MPC to a fresh tube.

21. To the recovered supernatant (step 20), add the equal volume (195 µl) of TE-saturated phenol/chloroform/isoamylalcohol, vortex thoroughly, and centrifuge at room temperature for 5 min at $14\,000 \times g$.

22. Recover 195 µl of upper aqueous layer to a fresh tube.

23. Add 2 µl of 1% linearized acrylamide, 19.5 µl of 3 M NaOAc, and 195 µl of isopropanol. Vortex thoroughly and centrifuge at 4 °C for 5 min at $14\,000 \times g$.

24. Discard the supernatant and add 200 µl of cold ethanol. Centrifuge at 4 °C for 1 min at $14\,000 \times g$.

25. Discard the supernatant. Dry the pellet at 37 °C for 5 min to evaporate residual ethanol.

26. Dissolve the pellet with 5 µl of TE.

27. Add 1 µl of 6× BPB dye (cDNA sample).

28. Apply cDNA sample to RLCS (1D gel electrophoresis).

Note:

1. It is important to pretreat Dynabeads M-280 as described in 'Reagents'. Here, the labeling reaction of the cDNA digested by *Bam*HI and/or *Bgl*II is described. Other restriction enzymes are

also used as 'restriction enzyme A'. In this reaction it is important to select the appropriate kinds of ddNTPs and use a fresh batch of $[\alpha^{32}P]$-dXTP(6000 Ci/mmol).

2. For successful RLCS it is preferable that the cpm (counts per minute) of the supernatant in step 13 is two orders lower than that of the pellet in the same step, and that the cpm of the recovered supernantant in step 20 is 70–80% of that of the mixture in step 18. We usually measure the cpm of each mixture, supernantant, or pellet, using a Cerenkov quick counter.

2-D Gel Electrophoresis

Note: For details of 2-D gel electrophoresis, refer Chapter 3.

Preparation of 1-D Agarose Gel

Vertical disk gel: 1% Sea-Kem GTG agarose / 5% sucrose in 1 × 1-D electrophoresis buffer

Gel apparatus: Teflon tubing (60 cm in length, 2.4 mm inner diameter)

Electrophoresis

1. Apply sample (6 µl) to the gel
   1-D electrophoresis
   1 × 1-D electrophoresis buffer
   300 V, ca. 16 h
   (Stop electrophoresis when the marker bromophenol blue reaches ~50 cm.)

In Situ Restriction Enzyme Digestion

1. 1-D gel rod taken from Teflon tube.

2. Incubate with 2nd restriction enzyme (restriction enzyme B) buffer (room temperature, 10 min, 2×).

3. Incubate with 2nd restriction enzyme (restriction enzyme B) solution (1 U/µl), 37 °C for 2 h.

4. Incubate with 1× TBE buffer (room temperature, 10 min, 2×).

Vertical 2-D 6% Nondenatured Polyacrylamide Gel Electrophoresis

1. Prepare 2-D 6% polyacrylamide gel (1× TBE buffer).

2. Transfer the 1-D gel rod to the top of 2-D gel, and connect both gels with melted 1% Sea-Kem GTG agarose in 1× TBE buffer (ca. 65 °C).

3. Apply 1× BPB dye on top of connecting gel.

4. Electrophoresis
   1× TBE buffer
   180 V, about 16 h
   (Stop electrophoresis when the marker bromophenol blue reaches ~35 cm)

5. Dry the gel at 65°C.

Autoradiography with intensifying screen at −80 °C for 3 days–1 week

Choice of Restriction Enzymes for RLCS

In RLCS, restriction enzyme reactions are carried out in two steps as shown in Fig. 8.1A and B; restriction enzymes A and B are used in enzyme digestion after cDNA synthesis reaction (Fig. 8.1A, step c) and in in situ gel digestion after the first-dimension gel electrophoresis (Fig. 8.1B), respectively. Some helpful hints on how to choose the enzymes used as enzyme A or B are described below.

Enzyme A. As described in Section 8.2, the enzyme which creates protruding cohesive 5′-termini should be chosen as enzyme A, because such termini are required for radiolabeling by Sequenase (Fig. 8.1A, step d) and are also useful for spot cloning (Fig. 8.4, Sect. 8.3.3). The number of spots which are displayed on one gel should also be considered for the choice of enzymes. Because approximately 1000 to 1200 spots seem to be appropriate for one gel, the use of six basecutters is recommended as enzyme A. This has been supported experimentally as follows. In mouse brain RLCS, when both six basecutters BamHI and BglII were used as enzyme A together and HinfI as enzyme B, about 1200 spots were displayed in one RLCS pattern. A similar spot number was also suggested in searching mouse brain cDNA data base on 100 cDNA sequences randomly selected. As enzyme A, which gives a similar spot number (900 to 1200 spots) when using *Hin*fI as enzyme B, an enzyme such as *Bln*I, *Bsp*HI, *Eco*RI, *Hin*dIII, or *Nco*I was selected by searching brain cDNA data base, and the number was also observed experimentally. Thus, the enzyme A may be chosen in a manner similar to that described.

Enzyme B. Enzyme B should be chosen as the enzyme whose sites are more, but not too many, in number compared with enzyme A sites in many kinds of cDNAs. Empirically, the five basecutters are preferable in this sense. The use of *Hin*fI (GANTC) or *Dde*I (CTNAG) may be recommended, since they generate an appropriate number of gel spots and are not expensive. Some four basecutters and six basecutters can also be utilized as enzyme B, if enzyme B generates an appropriate number of gel spots in combination with enzyme A.

Addition of ddNTPs in the Labeling Reaction

In the blocking reaction step (Fig. 8.1Ab) ddNTPs are added to reduce background signals as indicated in the experimental flow chart. However, it is preferable that only a few kinds of ddNTPs are added in the labeling reaction step (Fig. 8.1Ad). However, the kinds of ddNTPs to be added should be chosen so that they do not disturb ligation at the isotopically labeled restriction sites in the later cloning step. For example, *Bam*HI as enzyme A generates a 5′-protruding GATC sequence. In this case, ddC and ddT as well as labeled dG can be added, but neither ddG nor ddA should be added. Addition of ddG inhibits the incorporation of labeled dG, and elongation at the 3′-end of labeled dG by addition of ddA might disturb the later ligation reaction.

8.3.3
Spot Cloning for RLCS

Spot cloning for RLCS is schematically represented in Fig. 8.4. It is basically performed by the PCR-mediated method described in Section 8.4.3.

- 10× kination buffer   **Reagents**

- 0.1 M DTT

- 0.1 M ATP

- T4 polynucleotide kinase

- phenol-chloroform

- 2-butanol

- 10 M ammonium acetate

- Cold ethanol

- Chromaspin TE-30 (CLONTECH)

- 10× *Ex Taq* buffer (appended to *Ex Taq* polymerase)

- dNTPs mixture (appended to *Ex Taq* polymerase)

- *Ex Taq* polymerase (TaKaRa Shuzo Co. LTD)

Sequences – Sequence for ligation adapters
OLIGO(1):   5′-CGCCAGGGTTTTCCCAGTCACGACX-3′   (for enzyme A adapter)
OLIGO(2): 5′-XXXXGTCGTGACTGGGAAAACCCTGGCG (for enzyme A adapter)
OLIGO(3):   5′-CGCCAGGGTTTTCCCAGTCACGACY-3′   (for enzyme B adapter)
OLIGO(4): 5′-YYYYGTCGTGACTGGGAAAACCCTGGCG (for enzyme B adapter)

- Sequence for PCR primers
OLIGO(1) and OLIGO(3)

Note: These sequences are basically designated as M13 forward primer sequences. X and Y refer to enzyme A and enzyme B restriction sites, respectively.

Protocol Adapter Preparation

1. Combine the following components and fill up to 150 µl with DW
   Adapter oligonucleotides [OLIGO(2) or OLIGO (4)]   (10 µg)

   - 10× kination buffer   15 µl

   - 0.1 M DTT   7 µl

   - 0.1 M ATP   2 µl

   - T4 polynucleotide kinase   5 µl

2. Incubate at 37 °C for 1 h.

3. Phenol-chloroform treatment.

4. Concentrate by addition of 2-butanol (150 μl).

5. Centrifuge at 15 000 rpm at 4 °C for 10 s.

6. Concentrate by addition of 2-butanol (150 μl).

7. Centrifuge at 15 000 rpm at 4 °C for 10 s.

8. Recover aqueous layer (about 60 μl).

9. Add complementary oligonucleotide [OLIGO(1) or OLIGO (3)] (10 μg).

10. Add 10 M ammonium acetate (final 4 M).

11. Add cold ethanol (2.5 vol.).

12. Keep at −80 °C for 30 min.

13. Centrifuge at 15 000 rpm at 4 °C for 20 min.

14. Dry ppt.

15. Dissolve in 1 mM Tris-Cl (pH 8)/0.1 mM EDTA Na2 (pH 8) (20 μl).

16. Incubate at 60 °C for 10 min.

17. Cool to room temperature.

18. Adjust the concentration of adapters to 52 μM.

Recovery of Spot DNA from Gel and Adapter Ligation

Refer to Section 8.4.3. Use adapters appropriate for RLCS as described above.

PCR Amplification After Adapter Ligation

Note: Appropriate conditions of PCR amplification are different in the thermal cyclers. We show here the conditions for Thermal Sequencer TSR-300 (IWAKI Inc., Japan), Omni Gene (Hybaid. Inc.), and DNA Engine (MJ Research, Inc.). In the following, 'mixture A' and 'cycling A' are for Thermal Sequencer TSR-300, and 'mixture B' and 'cycling B' are for Omni Gene and DNA Engine. If other thermal cyclers are used, appropriate conditions are selected.

1. Purify the ligation mixture with spun column Chromaspin TE-30.

2. Combine the following components in a 0.5-ml reaction tube:

Mixture A
eluted ligation mix	$X \mu l$ (see below)
10× *Ex Taq* buffer	$5 \mu l$
dNTPs mixture	$4 \mu l$
OLIGO(1) (10 pmol/ml)	$1 \mu l$
OLIGO(3) (10 pmol/ml)	$1 \mu l$
DW	$(38-X) \mu l$
Ex Taq (5 U/ml)	$0.5 \mu l$

Mixture B
eluted ligation mix	$X \mu l$ (see below)
10× *Ex Taq* buffer	$2 \mu l$
dNTPs mixture	$1.6 \mu l$
OLIGO(1) (10 pmol/ml)	$1 \mu l$
OLIGO(3) (10 pmol/ml)	$1 \mu l$
DW	$(14.2-X) \mu l$
Ex Taq (5 U/ml)	$0.2 \mu l$

3. Overlay the reaction mixture with one drop of mineral oil.

4. Thermal cycling is performed as follows:

N cycles. The optimal number (N) of PCR cycles is determined by the intensity of spot signal and template volume (X). Under our standard condition, N is 25~35 and X is 1/10~1/20 volume of purified ligation mix. If you fail to amplify, increase X or N.

Cycling A   N cycles (see below):
94 °C 1 min
60 °C 1.5 min
72 °C 2 min
1 cycle
72 °C 5 min

Cycling B   1 cycle
98 °C	15 s
N cycles (see above):	
98 °C	15 s
65 °C	2 s

74 °C   35 s
1 cycle
72 °C   15 s

Note: This is the condition controlling the temparature of the heat block. If you use Omni Gene, set the parameter 'time ramping' that controls the changes in temperature in samples to 0.1 s/°C.

Cloning of Amplified Spot DNA Fragments

The PCR product is recovered with phenol/chloroform followed by ethanol precipitation and digested with the enzyme A used in the RLCS. The DNA fragments are purified using 6% nondenatured polyacrylamide gel electrophoresis. These DNA fragments have the 5′-protruding enzyme A site at the one end and a 3′-dA protruding end at the other. Thus, the DNA fragments are ligated to the dT vector with enzyme A site at one end (see Fig. 8.4), and transformed into *E.coli*. About 20 transformant colonies are usually screened with PCR using a vector primer set. The size of the amplified inserts and their restriction enzyme A sites should be examined. If the inserts in these clones are treated with four basecutters such as *Sau*3AI, clonal heterogeneity is checked by comparing their restriction site patterns with one another. If they are heterogenous, insert sizes and expression patterns are helpful to identify the required target clones.

8.4
Experimental Results and Discussion

8.4.1
RLCS Patterns

Poly(A)$^+$ RNA was prepared from tissues of adult mice. Several micrograms of poly(A)$^+$ RNA and 1 μg of the anchor primer (N = A) were used for cDNA synthesis. Under our experimental conditions, 1 μg of poly(A)$^+$ RNA, which was converted to 0.2–0.3 μg double-stranded cDNA with an estimated mean size of 1000 bp, was applied to gel electrophoresis. *Bam*HI and *Bgl*II were used as enzyme A, *Hin*fI as enzyme B.

Figure 8.3 shows RLCS patterns on cDNA samples prepared from mouse liver and brain (cerebral cortex, cerebellum, and brain stem). Many discrete spots with various intensities were

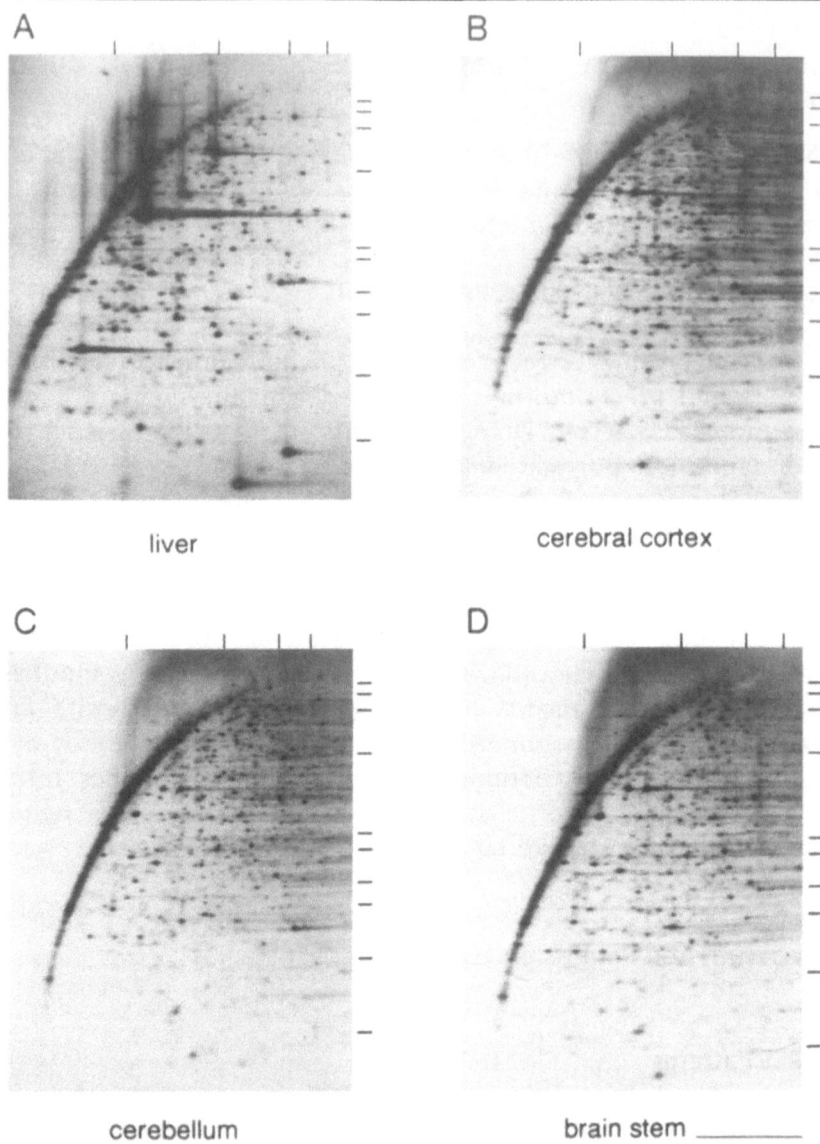

Fig. 8.3. Typical RLCS profiles for adult mouse liver (**A**), cerebral cortex (**B**), cerebellum (**C**), and brain stem (**D**). In each profile, *Bam*HI and *Bgl*II were used as restriction enzyme A, and *Hin*fl as restriction enzyme B. A λ-*Eco*T14I digest and a φX174-*Hae*III digest were used as the first- and second-dimension molecular markers, respectively. The scales in each figure are 1.88, 1.49, 0.93, and 0.42 kb (*from right to left*) and 1357, 1078, 872, 603, 310, 281/271, 234, 194, 118, and 72 bp (*from above to below*) for the first and second dimensions respectively. The *thick horizontal bar* is 10 cm. From H. Suzuki et al. [25] by permission of Oxford University Press

observed in the two-dimensional gels. We detected approximately 500 gel spots for the liver (Fig. 8.3A) and more than 1000 gel spots for each brain region (Fig. 8.3B, C and D). The spot patterns of the three brain regions were very similar, but were quite different from the pattern of the liver. Here we used an anchor primer with MA residues at the 3′-end, BamHI and BglII as restriction enzyme A, and HinfI as restriction enzyme B. If we use other anchor primers with MG, MC, or MT at the 3′-end and/or different sets of restriction enzymes, different spot patterns would be displayed, depending on the expression of other genes.

8.4.2
Correspondence Between Spot Intensity and Gene Expression Level

The spot patterns of the three brain regions were very similar but not identical. As shown in Fig. 8.5A, we could easily detect spots whose intensities were similar (spots S1 and S2) or different (spots S3, S4, and S5) among the three regions. We therefore examined whether the spot intensity correlates with the corresponding gene expression. cDNA fragments were isolated from the RLCS spots by the PCR-mediated spot cloning method (Fig. 8.4) and used as Northern blot hybridization probes. As indicated in Fig. 8.5B, Northern signals coincided well with the corresponding spot intensities. This means that the spot intensity reflects the corresponding gene expression level.

8.4.3
Sensitivity

Most (~90%) mRNA species are rare, each species being present at ~0.004% of total mRNA population of the cell [9]. The mRNA species that represent ~0.0001% of the total mRNA are classified as very rare mRNAs. In our RLCS experiment, 0.2–0.3 µg cDNA with an estimated mean size of 1000 bp was applied to a gel. Since the total number of the cDNA molecules is calculated to be 1.8–2.7×10^{11} copies, the copy number of cDNA species corresponding to very rare mRNA is calculated as 1.8–2.7×10^5 in the cDNA preparation used here. In RLGS, 1 µg of genomic DNA, equivalent to 3×10^5 copies, is used for a gel analysis and the main spots observed correspond to diploid intensity. Some spots are also detected with an intensity severalfold weaker than diploid spots. Thus, the spots with at least 10^5 copies of labeled DNA fragments can be detected

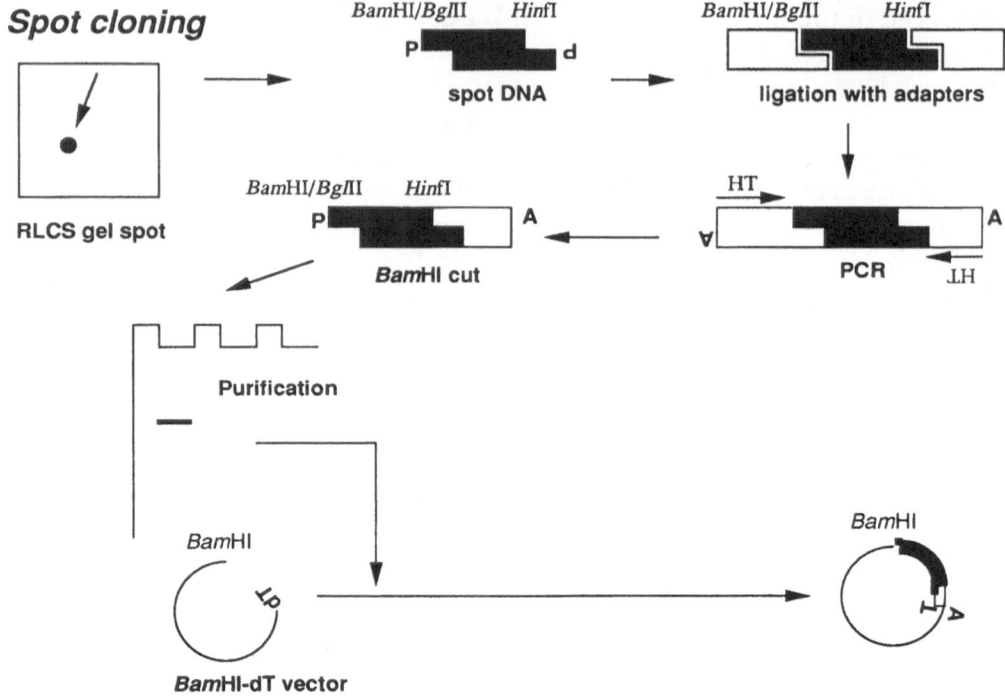

Fig. 8.4. Schematic representation of a PCR-mediated method for cloning spot DNA on RLCS gel

efficiently. This means that even very rare mRNA species are detectable as two-dimensional gel spots.

8.5
Perspective

In a new method for cDNA display, RLCS, described here, we can detect several hundred spots simultaneously. Since each spot and its intensity correspond essentially to one mRNA species, and its expression level, respectively, expression of many genes can be observed in an RLCS analysis. The expression of different genes is also displayed sequentially using different anchor primers with MN at the 3'-end and/or different sets of restriction enzymes. Furthermore, the method is sensitive enough to detect not only rare but also very rare mRNA species. As, however, most mRNA species are rare, it may be possible to display almost all the expressed genes with such modifications that nonoverlapping cDNA species are displayed as full-length cDNA. Recently, a modified

a

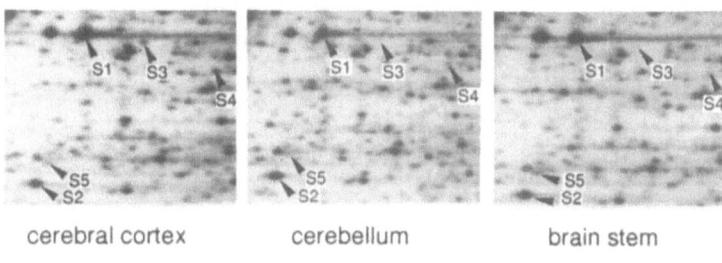

cerebral cortex   cerebellum   brain stem

b

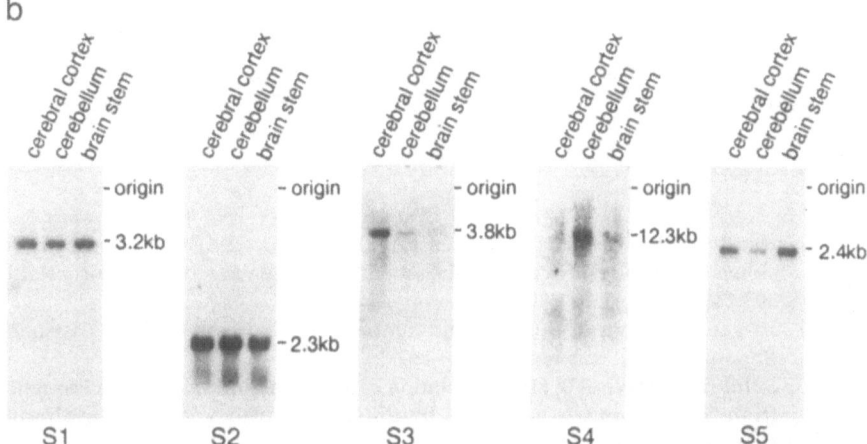

Fig. 8.5a,b. Correlation between the difference in intensity of the spots among RLCS profiles and expression level of the corresponding mRNA species among RNA samples used for the RLCS. **A** Profiles of the cloned spots. A part of the RLCS profiles is from Fig. 3B–D. The spots cloned in the present study (*S1–S5*) are shown by *arrowheads*. The *thick horizontal bar* is 5 cm. **B** RNA blot hybridization using cloned spot DNA fragments as probes. Two µg of poly (A)$^+$ RNA from the three brain regions were used for each lane. Cloned spot DNA fragments used as probes are described under the photographs. Each signal size described was estimated by using a 0.24–9.5 kb RNA ladder (Gibco BRL). The signal size of 12.3 kb in S4 was estimated by extrapolation. From H. Suzuki et al. [25] by permission of Oxford University Press

technique has been developed in differential display to minimize the redundancy which occurs on using arbitrary primers [27]. The method for rapid cloning of full-length cDNA is also being modified [28], and will be applicable to RLCS with further modifications. When such improvements are made in RLCS, it will in the future become a powerful technique to construct a transcript map of the genome as well as to analyze systematically differentially expressed genes and contribute to human genome research and other related projects.

8.6
Conclusion

We have developed a powerful and sensitive method, restriction landmark cDNA scanning (RLCS), which enables us to visualize many cDNA species simultaneously as two-dimensional gel spots. The method is also quantitative, since the intensity of the spots reflects the expression level of their corresponding genes. Thus, RLCS will be advantageous in the construction of transcript maps as well as in identification of differentially expressed genes.

References

1. Hendrick SM, Cohen DI, Nielsen EA, Davis MM (1984) Isolation of cDNA clones encoding T cell-specific membrane-associated proteins. Nature 308:149–153
2. Sargent TD, Dawid IB (1983) Differential gene expression in the gastrula of Xenopus laevis. Science 222:135–139
3. St John TP, Davis RW (1979) Isolation of galactose-inducible DNA sequences from Saccharomyces cerevisiae by differential plaque filter hybridization. Cell 16:443–452
4. Liang P, Pardee AB (1992) Differential display of eukaryotic messenger RNA by means of the polymerase chain reaction. Science 257:967–971
5. Bauer D, Müller H, Reich J, Ahrenkiel V, Warthoe P, Strauss M (1993) Identification of differentially expressed mRNA species by an improved display technique (DDRT-PCR). Nucleic Acids Res 21:4272–4280
6. Guimarães MJ, Lee F, Zlotnik A, McClanahan T (1995) Differential display by PCR: novel findings and applications. Nucleic Acids Res 23:1832–1833
7. Liang P, Averboukh L, Pardee AB (1993) Distribution and cloning of eukaryotic mRNAs by means of differential display. Nucleic Acids Res 21:3269–3275
8. Mou L, Miller H, Li J, Wang E, Chalifour L (1994) Improvements to the differential display method for gene analysis. Biochem Biophys Res Commun 199:564–569
9. Bertioli DJ, Schlichter UHA, Adams MJ, Burrows PR, Steinbiß H-H, Antoniw JF (1995) An analysis of differential display shows a strong bias towards high copy number mRNAs. Nucleic Acids Res 23:4520–4523
10. Hatada I, Hayashizaki Y, Hirotsune S, Komatsubara H, Mukai T (1991) A genomic scanning method for higher organisms using restriction sites as landmark. Proc Natl Acad Sci USA 88:9523–9527

11. Hayashizaki Y, Hirotsune S, Okazaki Y, Hatada I, Shibata H, Kawai J, Hirose K, Watanabe S, Fushiki S, Wada S, Sugimoto T, Kobayakawa K, Kawara T, Katsuki M, Shibuya T, Mukai T (1993) Restriction landmark genomic scanning method and its various applications. Electrophoresis 14:251–258

12. Hayashizaki Y, Hirotsune S, Okazaki Y, Shibata H, Akasako A, Muramatsu M, Kawai J, Hirasawa T, Shiroishi T, Watanabe S, Shiroishi T, Moriwaki K, Taylor BA, Matsuda Y, Elliott RW, Manly KF, Chapman VM (1994) A genetic linkage map of the mouse using restriction landmark genomic scanning (RLGS). Genetics 138:1207–1238

13. Okazaki Y, Hirose K, Hirotsune S, Okuizumi H, Sasaki N, Ohsumi T, Yoshiki A, Kusakabe M, Muramatsu M, Kawai J, Watanabe S, Plass C, Chapman VM, Nakao K, Katsuki M, Hayashizaki Y (1995) Direct detection and isolation of RLGS spot DNA marker tightly linked to a specific trait using RLGS spot-bombing method. Proc Natl Acad Sci USA 92:5610–5614

14. Kawai J, Hirose K, Fushiki S, Hirotsune S, Ozawa N, Hara A, Hayashizaki Y, Watanabe S (1994) Comparison of DNA methylation patterns among mouse cell lines by restriction landmark genomic scanning (RLGS). Mol Cell Biol 14:7421–7427

15. Kawai J, Suzuki H, Taga C, Hara A, Watanabe S (1995) Correspondence of RLGS-M spot behavior with tissue expression on mouse homologue of DP1/TB2 gene. Biochem Biophys Res Commun 213:967–974

16. Hayashizaki Y, Shibata H, Hirotsune S, Sugino H, Okazaki Y, Sasaki N, Hirose K, Imoto H, Okuizumi H, Muramatsu M, Komatsubara H, Shiroishi T, Moriwaki K, Katsuki M, Hatano N, Sasaki H, Ueda T, Mise N, Takagi N, Plass C, Chapman VM (1994) Identification of an imprinted U2af binding protein related sequence on mouse choromosome 11 using the RLGS method. Nat Genet 6:33–40

17. Shibata H, Yoshino K, Muramatsu M, Plass C, Chapman VM, Hayashizaki Y (1995) The use of restriction landmark genomic scanning to scan the mouse genome for endogenous loci with imprinted patterns of methylation. Electrophoresis 16:210–217

18. Watanabe S, Kawai J, Hirotsune S, Suzuki H, Hirose K, Taga C, Ozawa N, Fushiki S, Hayashizaki Y (1995) Accessibility to tissue-specific genes from methylation profiles of mouse brain genomic DNA. Electrophoresis 16:218–226

19. Kawai J, Hirotsune S, Hirose K, Fushiki S, Watanabe S, Hayashizaki Y (1993) Methylation profiles of genomic DNA of mouse developmental brain detected by restriction landmark genomic scanning (RLGS) method. Nucleic Acids Res 21:5604–5608

20. Hirotsune S, Hatada I, Komatsubara H, Nagai H, Kuma K, Kobayakawa K, Kawara T, Nakagawara A, Fujii K, Mukai T, Hayashizaki Y (1992) New spproach for detection of amplification in cancer DNA using restriction landmark genomic scanning. Cancer Res 52:3642–3647

21. Nagai H, Ponglikitmongkol M, Mita E, Ohmachi Y, Yoshikawa H, Saeki R, Yumoto Y, Nakanishi T, Matsubara K (1994) Aberation of genomic DNA in association with human hepatocellular carcinomas detected by 2-dimensional gel analysis. Cancer Res 54: 1545–1550

22. Miwa W, Yashima K, Sekine T, Sekiya T (1995) Demethylation of a repetitive DNA sequence in human cancers. Electrophoresis 16:227–232

23. Hirotsune S, Shibata H, Okazaki Y, Sugino H, Imoto H, Sasaki N, Hirose K, Okuizumi H, Muramatsu M, Plass C, Chapman VM, Tamatsukuri S,

Miyamoto C, Furuichi Y, Hayashizaki Y (1993) Molecular cloning of polymorphic markers on RLGS gel using the spot target cloning method. Biochem Biophys Res Commun 194:1406–1412

24. Suzuki H, Kawai J, Taga C, Ozawa N, Watanabe S (1994) A PCR-mediated method for cloning spot DNA on restriction landmark genomic scanning (RLGS) gel. DNA Res 1:175–180

25. Suzuki H, Yaoi T, Kawai J, Hara A, Kuwajima G, Watanabe S (1996) Restriction landmark cDNA scanning (RLCS): a novel cDNA display system using two-dimensional gel electrophoresis. Nucleic Acids Res 24:289–294

26. Chomczynski P, Sacchi N (1987) Single-step method of RNA isolation by acid guanidinium thiocyanate-phenol-chloroform extraction. Anal Biochem 162:156–159

27. Kato K (1995) Description of the entire mRNA population by a 3′ end cDNA fragment generated by class IIS restriction enzymes. Nucleic Acids Res 23:3685–3690

28. Schaefer BC (1995) Revolutions in rapid amplification of cDNA ends: new strategies for polymerase chain reaction cloning of full-length cDNA ends. Anal Biochem 227:255–273

Chapter 9

Concluding Remarks

Yoshihide Hayashizaki

The concept of genome scanning arose from the necessity to survey the genome with a biochemical approach to DNA. The conventional detection systems, Southern blotting and PCR, are mainly based on the specificity of DNA-DNA hybridization (Table 1.1). Southern blotting is the typical method which employs DNA-DNA hybridization between probe and DNA fragments immobilized on a filter [1,2]. PCR also adopts the hybridization specificity between the amplified DNA and two DNA primers whose sequences are designed to be the 5′ ends of the amplified DNA [3,4]. From this standpoint, RLGS and RLCS are classified as a different type of landmark-visualizing system, based on the new concept that restriction enzyme sites can be used as landmarks, resulting in various characteristic features and advantages.

Genome scanning is a typical top-down approach to analyzing the entire genome. The present goal in refining the technique is to improve the resolution of genome scanning in order to reach an analyzable level for the bottom-up approach. The resolution of genome scanning by RLGS has just been improved to under 1 Mbp in mammalian DNA. Higher resolution of genome scanning is required to reach a distance which can be covered by YAC, BAC, or PAC clones with no chimerism. The development of a new generation of genome scanning techniques with much higher resolution will enable us to analyze a much smaller distance that can be covered by one path sequencing. The RLGS method introduced here is the first generation RLGS (Ver. 1.8), and further development is underway in pursuit of greater numbers of landmarks in one assay and improved efficacy. In the future, new technologies for genomic scanning should aim for more information on the sequence and modification of genomic DNA.

References

1. Uitterlinden AG, Slagboom PE, Knook DL, Vijg J (1989) Two-dimensional DNA fingerprinting of human individuals. Proc Natl Acad Sci USA 86:2742–2746
2. Brilliant MH, Gondo Y, Eicher EM (1991) Direct molecular identification of the mouse pink-eyed unstable mutation by genome scanning. Science 252:566–569
3. Nelson DL, Ledbetter SA, Corbo L, Victoria MF, Ramirez-Slis R, Webster TD, Ledbetter DH, Caskey CT (1989) Alu polymerase chain reaction: a method for rapid isolation of human-specific sequences from complex DNA sources. Proc Natl Acad Sci USA 86:6686–6690
4. Dietrich W, Katz H, Lincoln SE, Shin H-S, Friedman J, Dracopoli NC, Lander ES (1992) A genetic map of the mouse suitable for typing intraspecific crosses. Genetics 131:423–447

Acknowledgments

The contributors to the development of the RLGS system are:

Yoshihide Hayashizaki	Chief Scientist, Genome Science Laboratory, Riken Tsukuba Life Science Center; Project Leader, RIKEN Genome Exploration Research Project
Yasushi Okazaki	Genome Science Laboratory, Riken Tsukuba Life Science Center
Shinji Hirotsune	Genome Science Laboratory, Riken Tsukuba Life Science Center
Nobuya Sasaki	Genome Science Laboratory, Riken Tsukuba Life Science Center
Hideo Shibata	Genome Science Laboratory, Riken Tsukuba Life Science Center
Hirosumi Imoto	Genome Science Laboratory, Riken Tsukuba Life Science Center
Hisato Okuizumi	Genome Science Laboratory, Riken Tsukuba Life Science Center
Tomoya Ohsumi	Genome Science Laboratory, Riken Tsukuba Life Science Center
Masami Muramatsu	Professor, Saitama Medical School; Consultant, Riken Tsukuba Life Science Center
Hideyuki Komatsubara	Toyobo Co., Ltd.
Sachihiko Watanabe	Senior Scientist, Shionogi Institute for Medical Science Osaka
Jun Kawai	Shionogi Institute for Medical Science Osaka
Harukazu Suzuki	Shionogi Institute for Medical Science Osaka
Takeshi Yaoi	Shionogi Institute for Medical Science Osaka

Verne M. Chapman	Roswell Park Cancer Institute in Buffalo, NY, USA
William A. Held	Roswell Park Cancer Institute in Buffalo, NY, USA
Christoph Plass	Roswell Park Cancer Institute in Buffalo, NY, USA
Kenneth F. Manly	Roswell Park Cancer Institute in Buffalo, NY, USA
Rosemary W. Elliott	Roswell Park Cancer Institute in Buffalo, NY, USA

The editors and authors of this book thank:

Kimi Akama	Genome Science Laboratory, Riken Tsukuba Life Science Center
Yukari Shigemoto	Genome Science Laboratory, Riken Tsukuba Life Science Center
Shigeko Tanaka	Genome Science Laboratory, Riken Tsukuba Life Science Center
Chie Owa	Genome Science Laboratory, Riken Tsukuba Life Science Center
Atsuko Akasako	National Cardiovascular Center
Keiko Matsui	National Cardiovascular Center
Arthur N. Westover	Genome Science Laboratory, Riken Tsukuba Life Science Center
Rieko Naka	Genome Science Laboratory, Riken Tsukuba Life Science Center
Masako Ohta	Genome Science Laboratory, Riken Tsukuba Life Science Center
Fuyuko Yoshida	Genome Science Laboratory, Riken Tsukuba Life Science Center
Rumiko Ebine	Genome Science Laboratory, Riken Tsukuba Life Science Center
Yousuke Mizuno	Genome Science Laboratory, Riken Tsukuba Life Science Center
Takeshi Hanami	Genome Science Laboratory, Riken Tsukuba Life Science Center
Mamoru Kamiya	Genome Science Laboratory, Riken Tsukuba Life Science Center
Takashi Sugimura	President, Emeritus at National Cancer Center; President, Toho University School

Acknowledgments 161

Takao Sekiya	Director, Department of Molecular Oncology at National Cancer Center
Nobutaka Takahashi	Director, Institute of Physical and Chemical Research
Kenichi Matsubara	Professor, Institute for Molecular and Cellular Biology Osaka University
Motoya Katsuki	Professor, Institute of Medical Science, University of Tokyo, Department of DNA Biology and Embryo Engineering
Akira Kira	Director, Riken Tsukuba Life Science Center
Tsutomu Hirasawa	Shionogi Institute for Medical Science Osaka
Kenji Hirose	Shionogi Institute for Medical Science Osaka
Chiharu Taga	Shionogi Institute for Medical Science Osaka
Ayako Hara	Shionogi Institute for Medical Science Osaka
Goro Kuwajima	Shionogi Institute for Medical Science Osaka
Yasuhiro Furuichi	Director, Agene Research Institute Co., Ltd.
Susumu Nishimura	Director, Banyu Pharmaceutical Co., Ltd.
Masaji Ohno	Director, Eisai Co., Ltd.
Susumu Tagaya	Eisai Co., Ltd.
Kazuo Moriwaki	Professor Emeritus, National Institute of Genetics
Toshihiko Shiroishi	Assistant Professor, National Institute of Genetics
Masahiko Nishimura	Assistant Professor, Institute for Expermental Animals Hamamatsu University School of Medicine
Masayuki Mori	Genome Science Laboratory, Riken Tsukuba Life Science Center; Institute for Experimental Animals Hamamatsu University School of Medicine
Shingo Akiyoshi	Genome Science Laboratory, Riken Tsukuba Life Science Center
Youichi Matsuda	Assistant Professor, Nagoya University
Norifumi Hirota	Iwasaki Electric Co., Ltd.

Masaaki Kitajima	Manager, New Business Development Department Japan Synthetic Rubber Co., Ltd.
Fan Kejun	New Business Development Department Japan Synthetic Rubber Co., Ltd.
Hiroyuki Tano	New Business Development Department Japan Synthetic Rubber Co., Ltd.
Junro Kuromitsu	Agene Research Institute Co., Ltd.
Hiroshi Kataoka	Agene Research Institute Co., Ltd.
Hideji Yamashita	Genome Science Laboratory, Riken Tsukuba Life Science Center, for their help.

Yoshihide Hayashizaki

Yasushi Okazaki

Shinji Hirotsune

Nobuya Sasaki

Hideo Shibata

Hirosumi Imoto

Hisato Okuizumi

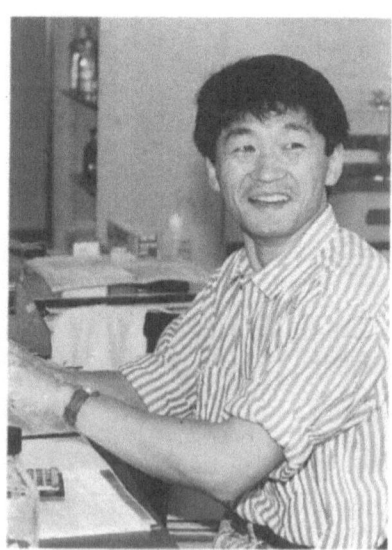

Tomoya Ohsumi

Masami Muramatsu

Hideyuki Komatsubara

Sachihiko Watanabe

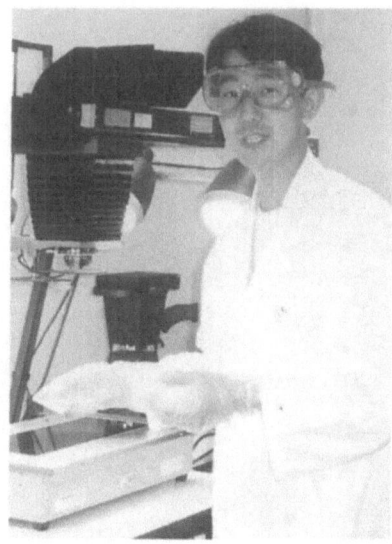

Jun Kawai

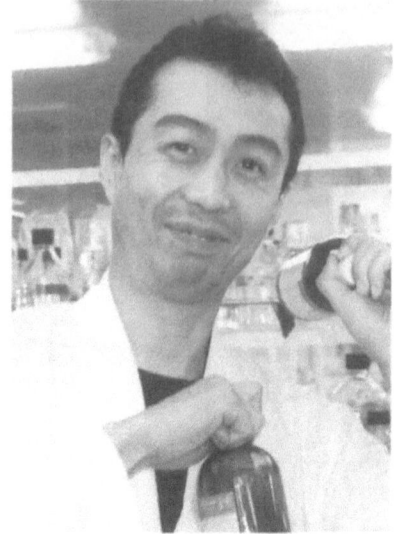

Harukazu Suzuki

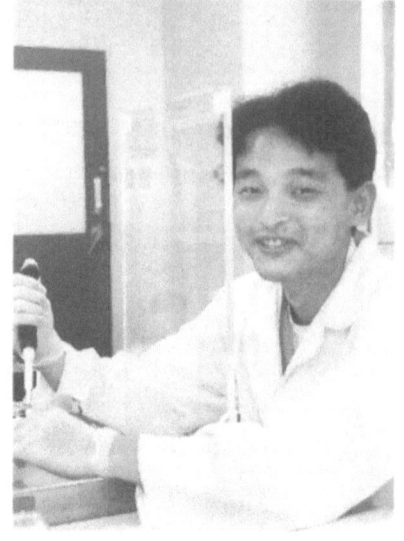

Takeshi Yaoi

Verne M. Chapman

William A. Held

Christoph Plass

Rosemary W. Elliott

Kenneth F. Manly

Abbreviations

BAC bacterial artificial chromosome

BPB bromophenol blue

BSA bovine serum albumin

DEPC diethyl pyrocarbonate

DTT dithiothreitol

EDTA ethylenediaminetetraacetic acid

FISH fluorescence in situ hybridization

HB high buffer

HCC hepatocellular carcinomas

LB low buffer

LN2 liquid nitrogen

LOH losses of heterozygosity

MB medium buffer

MPC magnetic particle concentrator

PEG polyethylene glycol

pBSII pBlueScript II

PCR polymerase chain reaction

PDP progeny distribution pattern

PFGE pulse field gel electrophoresis

RAPD random amplified polymorphic DNA

RFLP restriction fragment length polymorphism

RI recombinant inbred

RLCS restriction landmark cDNA scanning

RLGS restriction landmark genomic scanning

RLGS-M RLGS with methylation sensitive endonuclease

RLGS-SB RLGS spot bombing

SDS sodium dodecyl sulfate

SDP strain distribution pattern

SHB super high buffer

SSLP simple sequence length polymorphism

STS   sequence tagged site

TEMED   tetramethylene ethylene deamine

XC   xylene cyanol

YAC   yeast artificial chromosome

List of Suppliers

Supplier	Product
AGENE Research Institute Co., Ltd. 200 Kajiwara, Kamakura Kanagawa 247, Japan Tel.: 81. 467. 46-9590 FAX: 81. 467. 46-5320	*Not*I restriction trapper (Gene Trapper R-*Not*I)
Amersham International PLC UK Amersham Place Little Chalfont, Bucks HP79NA, UK Tel.: 44. 1494. 544-330 FAX: 44. 1494. 544-444	$[\alpha\text{-}^{32}\text{P}]$ dGTP (3000 Ci/mmol) $[\alpha\text{-}^{32}\text{P}]$ dCTP (6000 Ci/mmol) $[\alpha\text{-}^{32}\text{P}]$ ddATP (5000 Ci/mmol) Sequenase Ver. 2.0
Becton Dickinson Labware 1 Becton Drive, Franklin Lakes NJ 07417-1886, USA Tel.: 1. 201. 847-4200 FAX: 1. 201. 847-4854	Falcon tubes
Biocraft Inc. Suntopia 8F 1-42-9 Itabashi, Itabashi Tokyo 173, Japan Tel.: 81. 3. 3964-6561 FAX: 81. 3. 3964-6443	1st-D electrophoretic tank 1st-D gel holder Silicon stopper 2nd-D electrophoretic tank 2nd-D electrophoretic plate 2nd-D electrophoretic spacer

Supplier	Product
Boehringer Mannheim GmbH Biochemica D-68298 Mannheim, Germany Tel.: 49. 0621. 759-8531 TAX: 49. 0621. 759-8611	RNase A
BRL (GIBCO-BRL) → **Life Technologies Inc.**	
CLONTECH Laboratories, Inc. 4030 Fabian Way, Palo Alto CA 94303-4607, USA Tel.: 1. 415. 424-8222, 800. 662-CLON, 800. 662-2566 FAX: 1. 415. 424-1352, 415. 424-1064	Chromaspin TE-30, -100
Daiichi Pure Chemicals 2-8-3 Nihonbashi, Chuo-ku Tokyo 103, Japan Tel.: 81. 3. 5820-9408 FAX: 81. 3. 5820-9409	Acrylamide
DYNAL, Inc. USA/CANADA 5 Delaware Drive, Lake Success NY 11042, USA Tel.: 1. 800. 638-9416 FAX: 1. 516. 326-3298	Dynabeads(dT)25 Dynabeads M-280 Streptavidine Dynal MPC
Eastman Kodak Company Rochester, NY 14650, USA Tel.: 1. 716. 724-4000	Kodak XAR 5
FMC Inc. 191 Thomaston Street, Rockland ME 04841, USA Tel.: 1. 800. 341-1574	Sea-Kem GTG agarose

Supplier	Product
Hybaid. Ltd. 111–113 Waldegrave Road Teddington, Middlesex TW 11 8LL England Tel.: 44. 181-614 1000 FAX: 44. 181-977 0170	Omni Gene
IWAKI, Inc. 1-50-1 Kouda, Funabashi Chiba 273, Japan Tel.: 81. 474. 21-2090	Thermal Sequencer TSR-300
Japan Synthetic Rubber Co., Ltd. New Business Development Dept. 2-11-24 Tsukiji Chuo-ku Tokyo 104, Japan Tel.: 81. 3. 5565-6610 FAX: 81. 3. 5565-6637	*Not*I restriction trapper (Gene Trapper R-*Not*I)
Kimberly-Clark Crop. 1-12-22 Tsukiji, Chuo-ku Tokyo 104, Japan Tel.: 81. 3. 3543-2363 FAX: 81. 3. 3543-2814	Kimwipe

Kodak → Eastman Kodak Company

Supplier	Product
Life Technologies Inc. Gaithersburg, MD, USA Tel.: 1. 301. 840-8000 FAX: 1. 800. 331-2286	Sephacryl S-300 NACS column SUPERSCRIPT Choice System SuperScriptII RNaseH⁻ reverse transcriptase *E. coli* DNA ligase DNA polymerase I *E. coli* RNase H
Merck, E. Postfach 4119, Frankfurter Str. 250 W 6100 Darmstadt 1, FRG. Tel.: 49. 6151. 72-0 FAX: 49. 6151. 72-7521	Proteinase K
MJ Research, Inc. 149 Grove Street, Watertown MA 02172, USA Tel.: 1. 800. 729-2165 FAX: 1. 617. 923-8080	DNA Engine
Nakalai Tesque Inc. Nijo Karasuma, Nakagyo-ku Kyoto 604, Japan Tel.: 81. 75. 211-2516 FAX: 81. 75. 231-2455	Methylenebisacrylamide ammonium persulfate Tetramethylene ethylene- diamine (TEMED)
New England Biolabs, Inc. 32 Tozer Road, Beverly MA 01915-5599, USA Tel.: 1. 508. 927-5054 FAX: 1. 508. 921-1350	*Pac*I
PerSeptive Biosystems, Inc. 500 Old Connecticat Path Framingham, MA 01701, USA Tel.: 1. 508. 383-7700 FAX: 1. 508. 383-7881	BioMag Oligo (dT)20

Supplier	Product
Promega Corp. 2800 Woods Hollow Road Madison WI 53711-5899, USA Tel.: 1. 608. 274-4330, 800. 356-9526 FAX: 1. 608. 277-2516	RNase ONE ribonuclease
Sigma Chemical Company P. O. box 14508, St. Louis MO 63178, USA Tel.: 1. 314. 771-5765 FAX: 1. 314. 771-5757	Fraction V bovine serum albumin (BSA)
Stratagene 11011 North Torrey Pines Road, La Jolla, CA 92037, USA Tel.: 1. 619. 535-5400 FAX: 1. 619. 535-0045	pBlueScript II (pBSII)

Toyobo Co., Ltd.
2-2-8 Dojimahama, Kita-ku
Osaka 530, Japan
Tel.: 81. 6. 348-3786
FAX: 81. 6. 348-3322
	Dideoxynucleoside triphos-
	   phates (ddGTP,ddATP,
	   ddTTP,ddCTP)
	*Pvu*II, DNA polymerase I
	deoxynucleoside[α -thio]
	triphosphates(dGTP[α]S, q
	dCTP[α]S)

Takara Shuzo Co., Ltd.
13F Hankyu-Terminal Bldg.
1-1-4 Shibata, Kita-ku
Osaka 530, Japan
Tel.: 81. 6. 374-1685
FAX: 81. 6. 374-5457
 	
2-15-10, Nihonbashi, Chuo-ku
Tokyo 103, Japan
Tel.: 81. 3. 3271-8553
FAX: 81. 3. 3271-7282
	*Not*I, *Sca*I, *Pst*I, *Eco*RV, *Nco*I,
	   *Bam*HI, *Bgl*II, *Bln*I, *Eco*RI,
	   *Hin*dIII, *Hin*fI
	T4 DNA ligase
	E. coli DNA ligase
	Taq DNA polymerase
	Ex Taq polymerase
	T4 polynucleotide kinase
	T4 DNA ligase

Supplier	Product
Taitec 2693-1 Nishikata Koshigaya Saitama 343, Japan Tel.: 81. 489. 88-8347 FAX: 81. 489. 88-8350	Rotaton RT-50
United States Biochemical (USB) → **Amersham International PLC UK**	

Subject Index